THE NORTH AMERICAN GEESE

Their Biology and Behavior

The North American

Geese

Their Biology and

Behavior

Paul A. Johnsgard

School of Biological Sciences
University of Nebraska–Lincoln

Zea Books, Lincoln, Nebraska: 2016

Abstract

The eight currently recognized species of North American geese are part of a familiar group of birds collectively called waterfowl, all of which are smaller than swans and generally larger than ducks. They include the most popular of our aquatic gamebirds, with several million shot each year by sport hunters. Our two most abundant waterfowl, the Canada goose and snow goose, have populations collectively totaling about 15 million individuals. Like swans, the lifelong pair-bonding of geese, their familial care, and prolonged social attachment to their offspring are legendary. Their seasonal migratory flights sometimes span thousands of miles, and the sight of their long, wavering flight formations are as much the symbols of seasonal change as are the spring songs of cardinals or the appearance of autumnal leaf colors.

This book describes each species' geographic range and subspecies, its identification traits, weights and measurements, and criteria for its age and sex determination. Ecological and behavioral information includes each species' breeding and wintering habitats, its foods and foraging behavior, its local and long distance movements, and its relationships with other species. Reproductive information includes each species' age of maturity, pair-bond pattern, pair-forming behaviors, usual clutch sizes and incubation periods, brooding behavior, and postbreeding behavior. Mortality sources and rates of egg, young, and adult losses are also summarized, and each species' past and current North American populations are estimated. In addition to a text of nearly 60,000 words, the book includes 8 maps, 21 line drawings, and 28 photographs by the author, as well as more than 700 literature citations.

Text, photos, and illustrations copyright © 2016 Paul A. Johnsgard

ISBN 978-1-60962-094-3 paperback
ISBN 978-1-60962-095-0 ebook

Composed in Adobe Garamond and Imprint MT Shadow types.

Zea Books are published by the University of Nebraska–Lincoln Libraries
Electronic (pdf) edition available online at http://digitalcommons.unl.edu
Print edition available from http://www.lulu.com/spotlight/unlib

UNL does not discriminate based upon any protected status. Please go to unl.edu/nondiscrimination

Contents

Maps

Figures

Photographs

Preface

The eight species of North American geese are part of an easily recognized group of birds that are collectively called waterfowl (wildfowl in Britain). They are typically smaller than swans and larger than ducks, and range in size from less than four pounds to about 16 pounds. They include the most popular of our gamebirds—several million are shot each year by sport hunters—and our most abundant waterfowl, the Canada goose and snow goose, whose world populations currently total about 20 million individuals.

Geese, together with swans, are part of a distinctive subfamily (Anserinae) of the waterfowl family of ducks, geese, and swans (Anatidae). Nearly all of the 15 living species of geese are moderately large to very large waterfowl and, in common with swans and whistling ducks (Dendrocygnini), have plumage patterns that are alike in both sexes and lack iridescent coloration. They also all similarly possess a weblike scale pattern on their lower leg surface unlike that of typical ducks and transitional forms such as sheldgeese. In geese, swans, and whistling ducks pair-bonds are usually permanent, and courtship behavior is correspondingly relatively simple and nearly identical in both sexes. Male assistance with rearing the young, and sometimes also with incubation, is characteristic.

Like most swans but unlike the tropically distributed whistling ducks, all 17 of the now-recognized species of geese are found in the cooler parts of the Northern Hemisphere, and most undertake fairly long migrations between their wintering areas and their more northern, often Arctic, breeding grounds. In conjunction with their long migrations, geese tend to be highly social, their flocks mostly consisting of assemblages of pairs and families. Family and pair bonds are maintained by complex vocal "signals" and visual devices that help to coordinate each social unit's movements and activities. Unlike most whistling ducks and many true ducks, goose vocalizations lack pure whistled notes and are notable for the harmonic components that are often present. Such vocal complexity allows acoustic diversity within species, and probably facilitates individual recognition among pairs and closely related birds.

All geese are primarily vegetarians, obtaining much of their food from terrestrial surface vegetation in the case of most geese, but sometimes from submerged rootstocks or, rarely, from animal sources. Adult goose plumages are alike in both sexes, lack iridescence, and are fairly simple with brown and white colors predominating.

Adult (definitive) plumages in geese are acquired by the end of the first year of life, although sexual maturity and initial breeding might not occur for several years. A single wing and body molt per year is typical of geese, so seasonal plumage variations do not exist, except for possible fading or incidental staining of feathers acquired in the course of foraging. In a few species, genetically controlled "morph" plumage variations occur, such as the white and "blue" plumage morphs of snow geese, and three different downy plumage morphs of Ross's geese.

Map 1. Approximate boundaries of Atlantic, Mississippi, and Pacific Flyways. To avoid confusion, limits of the Mississippi Flyway are indicated by connected dots. After Johnsgard, 2012.

Since the 1930s, the US Fish and Wildlife Service (previously the US Biological Survey) has interpreted waterfowl migrations largely in terms of "flyways," which are multistate and multiprovincial regions that were constructed on the basis of waterfowl banding data. The resulting four geographic constructs (the Atlantic, Mississippi, Central, and Pacific flyways) were invented to help define and illustrate major migration routes of ducks, geese, and swans in North America.

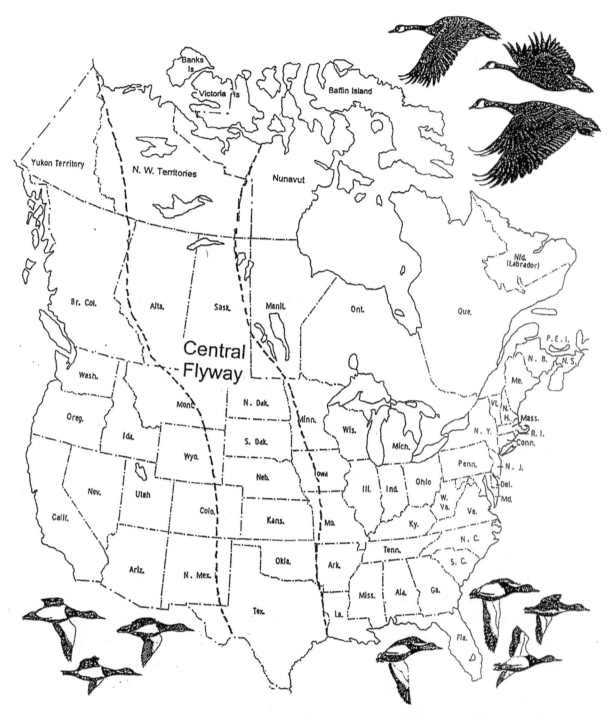

*Map 2. Approximate boundaries of Central Flyway. Note that unlike in the following species'
range maps, the geographic boundaries of the recently established Nunavut Territory are shown
here, replacing the previous Keewatin District and part of the Mackenzie District of Canada's
Northwest Territories.*

Although their utility in describing such routes for the nearly 50 species of North
American waterfowl is mostly limited to these and other water-dependent birds, the flyway concept has been used to assemble state and provincial agencies into administrative

units that are helpful in formulating hunting regulations and defining regional conservation issues. For example, breeding and winter waterfowl surveys are often organized on the basis of flyways, and annual hunter kills (euphemistically labeled "harvests" by government agencies) are also usually estimated using flyway categories. Because flyways are frequently referred to in the following text and related waterfowl literature, maps of their traditional limits are included in this book (maps 1 and 2).

Species-level nomenclature in this book follows current American Ornithologists' Union (AOU) usage, but generic categories follow my earlier books (*Chen* and *Philacte* are included within *Anser*). The taxonomic sequence of species that I have used incorporates my own taxonomic conclusions (1961, 1965) plus those of Livezey (1996). Unlike AOU practice, I have provided English names for subspecies to provide reading specificity and to simplify taxonomic language.

Recent changes in technical classification and names (biological nomenclature) used for North American geese have had an impact on both their popular terminology and their technical literature. An important recent change in goose classification was the taxonomic "splitting" in 2004 of the Canada goose into two species by the AOU. At that time, several populations that had previously been regarded as small races of Canada geese were distinguished from the Canada goose on the basis of their genetic differences and renamed the cackling goose. Because the two species are inseparable by simple plumage characteristics, this separation has complicated both the popular and technical literature of Canada and cackling geese and has made mapping their respective breeding and ranges nearly impossible. Almost all the maps from this book's original edition have been modified to varying degrees to try to incorporate new range information and actual distributional changes, but some of these changes now occur remarkably rapidly, and my current mapping efforts will soon be outdated.

Most of the North American geese have breeding and/or wintering ranges extending to Eurasia (greater white-fronted goose, brant) or at least to eastern Asia (snow goose, emperor goose). Additionally, the Canada goose is now well established as an introduced species in Europe, New Zealand, and elsewhere. The barnacle goose is included in this book as a North American goose mostly "by courtesy only" because it breeds no closer than eastern Greenland but is a fairly regular visitor to our Atlantic shores and occasionally occurs in the continental interior.

About half of the text that follows represents an updated version of two of my earlier writings: *Waterfowl of North America* and *Ducks, Geese, and Swans of the World.* Since these books were published in the 1970s, a vast number of research publications have been produced on North American geese, and I have tried to include documentation for as many of these newer sources as is feasible. I have included new text primarily where information in my earlier summaries was weak or lacking; where it seemed adequate I have kept the original accounts largely intact.

Interestingly, among a total of about 600 references that were in my 1975 *Waterfowl of North America*, single authors wrote 66 percent, and only 5 percent were by three or more authors. In contrast, among more than 700 references in this text, 35 percent were produced by from 3 to 25 authors, and only 30 percent were by single authors. This strong current trend toward multiple authorship probably reflects the increased number of researchers now cooperatively engaged in single large research projects, and the frequent use of technologically more diverse and highly skill-dependent tools.

As with my previous monographs already placed into the UNL DigitalCommons library, I owe a huge debt of gratitude to Paul Royster, Coordinator for Scholarly Communications for the University of Nebraska–Lincoln Libraries and publisher of Zea Books, for accepting and seeing this project through to completion, and to the UNL School of Biological Sciences for continuing to support my research.

Paul A. Johnsgard
Foundation Regents Professor Emeritus
Biological Sciences
University of Nebraska–Lincoln
July 2016

Lesser snow goose, adult landing

Barnacle goose, adult foraging

I.

Species Accounts

Emperor geese, adults in flight

Emperor Goose

Anser canagicus (Sewastianov) 1802 (*Chen canagica* of AOU, 1998)

Other vernacular names. Beach goose

Range. Breeds in coastal Alaska from the mouth of the Kuskokwim River to the north side of the Seward Peninsula, St. Lawrence Island, and on the northeastern coast of Siberia. Molting once mostly occurred on St. Lawrence Island but more recently has been on the Chukotski Peninsula. Winters on the Aleutian Islands and along the Alaska Peninsula to Kodiak Island, with vagrant birds wintering in British Columbia and the western United States south rarely to California.

Subspecies. None recognized.

Map 3. Breeding (hatched) and wintering (shaded) distributions of the emperor goose in North America.

Measurements. (*Note:* In this work, wing measurements are given as the distance from the bend of the wrist to the tip of the longest primary while the wing is folded and unflattened [chord distance], unless otherwise indicated.)

Folded wing: males 380–400 mm, females 350–385 mm (Delacour, 1954); ave. of 121 females 377.3 mm (Petersen, Schmutz, and Rockwell, 1994).

Culmen: Males 40–49 mm, females 35–40 mm (after Delacour, 1954): ave. of 86 females 35.7 mm (Petersen, Schmutz, and Rockwell, 1994).

Weights (mass). (*Note:* In this work, weight [mass] measurements were originally reported in pounds. Pounds [lb.] are now also shown with approximate gram [g] equivalents.)

Ave. of 6 males 6.2 lb. (2,812 g), max. 6.8 lb. (3,084 g); 9 females ave. 6.1 lb. (2,766 g), max. 6.9 lb. (3,130 g) (Nelson and Martin, 1953). Ave. of 126 females 1,638 g (Petersen, Schmutz, and Rockwell, 1994). Ave. of males (number unstated) 2,812 g, females 2,766 g (Madge and Burn, 1988).

Identification

In the hand. Emperor geese can hardly be confused with any other species when in the hand. The multicolored reddish bill lacking exposed "teeth," the yellowish to yellow-orange legs and feet, and a distinctive scalloped feather pattern of gray, black, and white are all unique.

In the field. Along their very limited range, emperor geese are usually found along saltwater shorelines, where they occur in small flocks. The golden to orange staining on their white head feathers is often conspicuous and contrasts with the otherwise grayish plumage. In flight, the lack of white feathers above or below the tail makes this species unique among geese. They also have relatively short necks and heavy bodies, associated with a rapid and strong wing beat. Flying birds often utter a repeated *kla-ha* or an alarm note *u-leegh*.

Age and Sex Criteria

Sex determination. No plumage characters are available for external sex determination.

Age determination. Brown, rather than black, barring on the back, brown secondaries, and wing-coverts, and gray mottling on the white head and neck indicate a first-year bird. The legs and feet of juveniles and immatures are dull olive black rather than the bright yellow-orange of adults. Breeding adults often have golden to orange stains on the white head and neck feathers from foraging in iron-rich or highly organic water, but these are usually lacking in immatures.

Emperor goose, adult portrait

Emperor goose, adult at rest

Distribution and Habitat

Breeding distribution and habitat. The emperor goose's breeding distribution in North America is one of the most restricted of our goose species, and is limited to the west coast and adjacent islands of Alaska. Gabrielson and Lincoln (1959) described the overall range as extending from Kotzebue on the north to the Aleutian Islands on the south, with the chief breeding occurring from the mouth of the Kuskokwim River to the north side of the Seward Peninsula. The most eastern breeding record is reported for Cape Barrow, where a pair was taken in 1929, and the most southerly is for Amak Island. It was uncertain to Gabrielson and Lincoln whether birds on St. Lawrence Island were nesters or simply nonbreeding and molting birds, but Fay (1961) established that both breeding and molting does occur there. Bailey (1948) found that emperor geese were common nesters on the north shore of the Seward Peninsula and thought they were probably less common nesters on the north shore of Kotzebue Sound to at least Point Hope.

Throughout their North American range, favored nesting habitats are in low, wet tundra, usually near the coast and often near lakes or ponds. Conover (1926) reported that nesting at Hooper Bay occurred within ten miles of the coast. Spencer et al. (1951) noted that although emperor geese nested in association with cackling geese 12 to 15 miles from the coast in this area, they also nested farther inland, with white-fronted geese and lesser Canada geese. Barry (1964) noted that ponds and marshes in low, rolling hills, inland from the tidal areas favored by brant, were preferred nesting habitat, with emperors, cackling geese, and brant geese overlapping somewhat in their nesting habitat zones.

In Russia (Siberia), nesting occurs over a broad area adjoining the Bering Sea, from Anadyr north and west to eastern Chukotski Sea, probably to Kolyuchin Bay, about 175°W longitude. Favored nesting habitats consist of coastal flats, islands in the mouths of small rivers emptying into the sea, and to some extent of swampy marshes along the lower reaches of rivers flowing through tundra (Dementiev and Gladkov, 1967). Kistchinski (1971) also reported that coastal "lagoon" tundra and inland moss-sedge tundra represented the species' two main nesting habitats.

Wintering distribution and habitat. Virtually the entire emperor goose population of North America is believed to winter along the Aleutian Islands. They also winter on the Commander (Komandorski) Islands, inhabiting stony, rubble-covered coasts (Dementiev and Gladkov, 1967).

In 1961 Kenyon estimated there were 25,000 to 37,000 wintering birds in the Aleutian Islands and added that since others probably wintered along the Alaska Peninsula, the total winter population might have been around 200,000 birds in about 1960. The birds were abundant in winter around Kanaga Island but had been reported all the way from the Sanak group to Attu (Murie, 1959). In some winters large numbers were seen at Izembek Bay. However, from 1964 to 1986 the Alaska population declined from 139,000 to 42,000 (Petersen, Schmutz, and Rockwell, 1994), and 91,900 were estimated in the spring of 2009 (Baldassarre, 2014).

In some winters emperor geese have been seen along the west coast from British Columbia to coastal southern California and even inland to Idaho. These sightings are believed to be the result of the transferring of some emperor goose eggs to the nests of white-fronted geese by wildlife biologists, with a resultant shift in wintering movements (*Audubon Field Notes* 24: 633). There are many recent eBird records from southeastern Alaska (e.g., Gustavus, Mendenhall wetlands), British Columbia (Victoria Island, Vancouver area), Washington (Dungeness, Snohomish, Seattle area), Oregon coast (Astoria to Brookings), Idaho (Boise), and California coast (south to Long Beach).

General Biology

Age at maturity. Ferguson (1966) reported that 15 of 17 aviculturists responding to a questionnaire indicated that initial breeding of captive emperor geese occurred when they were three years old, with the other two reporting two years and five years. Schmutz (2001a) found that, of 13 marked nesting females, three were three-year-olds, seven were four-year-olds, and three were five-year-olds. No birds younger than three years of age were found on nests.

Pair-bond pattern. Little documented information is available about pair-bond patterns, but most observers have noted strong pair-bonds, which are apparently permanent (Petersen, Schmutz, and Rockwell, 1994).

Emperor geese, adults at rest

Nest location. Nests are typically placed near water, such as on an island, a bank, or in a large tussock (Conover, 1926). Sometimes the driftwood debris on the high tide line is chosen for concealing the nest (Barry, 1964). In the Hooper Bay region we noted (Kessel et al., 1964) that 13 nests were all in grassy marsh habitat, mostly within a few feet of water but sometimes 20 to 40 feet from the nearest pond. Calvin Lensink (pers. comm.) reported that emperor geese nest farther from the coast than do brant and more often are found nesting along the main shoreline than on small islets.

Around Hooper Bay they often nest in upland hummocks or "pingos" several yards from water, and on other coastal flats they might nest in clumps of wild rye (*Elymus*) well away from water. Eisenhauer and Kirkpatrick (1977) reported that 260 nests in lowland pingo tundra at Kokechek Bay averaged spacing of 58.2 meters apart but were more scattered in other habitats. The pair defended an area of about 14 square meters around the nest during the period prior to egg laying.

Eisenhauer and Kirkpatrick also reported that, relative to random locations, emperor geese select sites with larger amounts of shrubs, farther from open water, near ponds with few islands, and located lower on the sides of pingos. Successful nests tend to be surrounded by tall, dead vegetation. In areas of dense nestings, sites used during previous years might be selected. Second nesting efforts have not been reported (Petersen, Schmutz, and Rockwell, 1994).

Clutch size. Of 500 active nests that were found on the Yukon-Kuskokwim Delta between 1963 and 1971, the clutches averaged 4.72 eggs, with yearly means ranging from 3.83 to 5.59 (Calvin Lensink, pers. comm.). This area perhaps supports as much as 90 percent of the world's emperor goose population and must represent optimum habitat, but in Siberia clutch sizes are comparable, usually of 5 to 6 eggs (Dementiev and Gladkov, 1967). The mean egg-laying rate for a five-egg clutch is about 6.3 days (Petersen, Schmutz, and Rockwell, 1994).

Nest and egg losses. Losses to egg predators, principally jaegers, reduced the average clutch from 5.5 to 3.8 in one study (*United States Department of the Interior Resource Publication* 43, p.19, 1967), or an approximate 30 percent egg loss. Brood counts made in 1950 and 1954 (*United States Fish and Wildlife Service, Special Scientific Report: Wildlife*, Nos. 8 and 27) indicate an average brood size of 4.5 for 28 broods, suggesting a somewhat low early mortality, assuming no brood mergers occurred. Calvin Lensink (pers. comm.) reported an average of 3.85 goslings in 318 early (Class I) broods.

Juvenile mortality. Fairly substantial losses of newly hatched goslings to glaucous gulls have been noted by various observers (Brandt, 1943; Conover, 1926). Arctic foxes have also been reported to prey on both eggs and young where they are abundant (Barry, 1964). Egg predators include jaegers and foxes, and glaucous gulls are primary predators on goslings. Bald eagles take both adults and juveniles (Petersen, Schmutz, and Rockwell, 1994).

Adult mortality. Petersen (1992c) judged a 60 percent survival rate of neck-collared adult females for 1982–1985, and Schmutz, Cantor, and Petersen (1994) used the same technique for estimating monthly over-winter survival (94 percent for all adults, 71 percent for all juveniles) and monthly over-summer survival (98 percent for adults, 94.3 percent for juveniles). The estimated annual survival rate of both sexes of adults was 63.1 percent. Annual estimations of juveniles in fall staging areas since 1985 produced an overall average of 19.1 percent juveniles, with a range of 9.2–35.2 percent, and an estimated overall annual survival rate of 83 percent (Dau et al., 2006). Hupp et al. (2008) calculated annual survival rates of radio-tagged adult females as 79–85 percent, with local subsistence hunting apparently a major contributing factor to the species' recent low population levels.

General Ecology

Food and foraging. The emperor goose has been aptly called the "beach goose" as a reflection of its littoral foraging tendencies. Cottam and Knappen (1939) provided much of the available data on the foods of this species. In their sample of 33 stomachs, mostly from spring and summer specimens from Alaska, the contents were almost entirely (91.6 percent) vegetable material. Only two of the birds had been feeding predominantly on animal material, a finding in contrast to most opinions on foraging tendencies of emperor geese. Major spring and summer food sources consisted of algae (30.7 percent), eelgrass and pondweeds (13.9 percent), grasses and sedges (24.9 percent), and unidentified plant fiber (22 percent).

Animal matter included mollusks (3.7 percent), crustaceans (2.2 percent), and other materials (2.6 percent). Sea lettuce (*Ulva* and *Enteromorpha*) made up 17 percent of the total and occurred in 12 stomachs, while the remainder of the algae consisted of green algae. Eelgrass (*Zostera*) is apparently also a favored food, judging from its occurrence in the samples (Cottam and Knappen, 1939). Barry (1964) noted that young birds feed on aquatic insects and marsh grass at first, and later might consume berries such as crowberries (*Empetrum*) (Barry, 1964). Late summer foods of molting birds consist largely of sedges and grasses (Kistchinski, 1971).

Murie (1959), in referring to wintering birds, commented on their use of kelp, sea lettuce, and *Elymus* shoots. Dementiev and Gladkov (1967) mentioned that various invertebrates, particularly mussels and other mollusks and crustaceans collected in the tidal zone, are major sources of food. Quite possibly there are local or seasonal variations in the dependence upon animal foods by this species. A high level of consumption of animal foods (mussels, clams, other invertebrates) by wintering emperor geese is unique among North American geese (Schmutz, 1994).

Sociality, densities, territoriality. Brandt (1943) noted that during spring migration the geese moved northward in flocks of about 15 to 40 birds and that early arrivals at the nesting grounds were in pairs or small groups. Arrivals on the Alaskan breeding grounds soon scatter over the available habitats and nest in what might be called loose colonies, with nests averaging about 200 feet apart and approximately one nest per four acres (Eisenhauer et al., 1971; Eisenhauer and Frazer, 1972). Little has been written on the prenesting behavior of the emperor goose. During early June, I noticed a total of 400 to 500 emperor geese within a few square miles of the Hooper/Igiak Bay lowland (Kessel et al., 1964), which were mostly in groups of no more than a few dozen birds.

Breeding densities over broad areas have not been carefully estimated, but in the Hooper Bay area the emperor goose makes up about 10 percent of the breeding waterfowl population, which has been estimated at 130 birds per square mile (Spencer et al., 1951), so a density of 6 to 7 pairs per square mile would be suggested. This compares well with a more recent estimate made by Mickelson (1973) of 20 pairs on a four-square-mile study area. There were also 204 cackling goose pairs, 32 black brant pairs, 19 white-fronted goose pairs, and 42 spectacled eider pairs present on that area. On Schmutz's (2001) larger study area in 1990, there were 399 cackling goose nests, 56 white-fronted goose nests, and 89 emperor goose nests, whereas in 1997 these respective numbers were 339, 93, and 46.

Interspecific relationships. Nesting is done in the same general habitat as is used by cackling geese and greater white-fronted geese, but suitable nest locations are probably never lacking in typical lowland tundra habitats. There should also seemingly be little if any food competition between emperor geese and

Emperor geese, two adults swimming

any other species of geese in lush lowland coastal tundra. Emperor goslings, however, preferentially eat sedges (especially *C. subspatheca*), as do the goslings of cackling geese and greater white-fronted geese. Because cackling geese are from two to five times more common than emperor geese on the Yukon-Kuskokwim Delta, interspecific competition for food might be strong during the brood-rearing period and can affect the development of emperor goose goslings (Lake et al., 2008).

General activity patterns and movements. During the long Arctic summer days on the nesting grounds at Hooper Bay, I observed no definite schedule of activities for the emperor geese. Nonbreeding birds or birds that were still in the process of egg laying could be seen foraging around the edges of tundra ponds at almost any hour, usually in pairs or what appeared to be family groups of 5 to 7 birds. They were far less wary and more "curious" than any of the other geese, and, when flushed, they would typically circle almost at eye level several times around the person flushing them before flying away. The Inuits thus found them easy targets and, even with a single-shot .22-caliber rifle could usually kill more than one bird from a flock before it finally left the area. Several thousand are probably killed annually on the Yukon-Kuskokwim Delta by subsistence hunters.

During early September at Izembek Bay, I have observed migrant birds foraging along the beaches in the tidal zone, and rarely if ever did they undertake daily flights to the tundra to feed on berries, as is typical of cackling geese.

Social and Sexual Behavior

Flocking behavior. As noted earlier, large flocks of emperor geese are rarely encountered, except perhaps in summer molting areas (Pay, 1961). A relatively large winter flock of 2,350 birds was seen during an Audubon Christmas Bird Count at Izembek Bay, Alaska (*Audubon Field Notes*, 22: 114).

Pair-forming behavior. Pair formation probably occurs at the wintering areas, since the birds arrive at their breeding areas already in pairs (Bailey, 1948; Brandt, 1943). I observed no pair-forming behavior at Hooper Bay and saw no aggressive behavior in the small groups that moved about together, suggesting that they were family units. My observations of geese in captivity indicate that a typical triumph ceremony is present, which no doubt serves to establish and maintain bonds in emperor geese, as in other goose species.

Copulatory behavior. I have never observed a completed copulation, and the only apparent precopulatory behavior I have seen rather closely resembled normal feeding behavior on the part of both birds. Brandt (1943) noted that mating occurred in shallow water, just deep enough to allow the female to sink beneath the surface. It is likely that most copulations occur before or during spring migration, as very few have been reported from the breeding grounds (Eisenhauer and Kirkpatrick, 1977).

Nesting and brooding behavior. According to Brandt, the female builds a nest in grasses usually close to water, first hollowing out a cup from 2.2 to 4.5 inches deep, enough to allow the female to be well concealed but also leaving an adequate accumulation of grasses and moss below. Incubation begins with the completion of the clutch, but little down is added until near the end of incubation, when it is liberally deposited. The male remains near the incubating female but not as close as in the greater white-fronted geese. Following hatching the male joins the family, and they move to rivers and sloughs near the coast where the young forage for aquatic insects or might feed on sedges (*Carex* spp.) and tundra berries with their parents.

The clutch size averages slightly less than 5 eggs, with some yearly variations that are evidently associated with prevailing weather and habitat conditions. The mean clutch size of 293 nests on the Yukon-Kuskokwim Delta was 4.8 eggs between 1963 and 1972 but ranged annually from 3.8 to 5.3 (Eisenhauer and Kirkpatrick, 1977). Only the female incubates, and the typical incubation period under wild conditions has been reported as 24 to 25 days, with extremes of 23 to 27 days. The modal incubation period at Kokechik Bay was 24 days, with the male typically standing close by and responding to predators (Eisenhauer and Kirkpatrick 1977).

A surprisingly high percentage of the nests on the Yukon-Kuskokwim Delta are parasitized by other emperor geese, or by cackling geese, greater white-fronted geese, or spectacled eiders (Watkins, 2006).

Emperor geese, two adults swimming

Such parasitism has little effect on the hatchability of host eggs, reducing hatching success in one study by only about 4.5 percent, from 93.2 to 88 percent, and nearly half of the parasitically laid eggs also hatched (Petersen et al., 1994).

Major egg predators would appear to be jaegers, although following hatching the young are taken by a variety of species, including glaucous and glaucous-winged gulls, three species of jaegers, and perhaps also the snowy owl (Brandt, 1943). Eisenhauer and Kirkpatrick (1977) found a very low incidence of egg mortality by foxes, with jaegers and glaucous gulls much more significant, although a later study (Petersen, 1992a) in the same region found Arctic foxes to be the most important egg predator. Gulls and jaegers are evidently the primary mortality factors for both eggs and young in Alaska; observations in Siberia by Kistchinski led him to believe that inadequate insulation of the nest and poor resistance of the goslings to cold might limit the northward extension of the species' range.

With the hatching of the eggs, the families usually move initially to the coast, where they might remain for a time before going to the larger rivers for molting. Molting areas for breeders might be as far as five miles from the nesting site (Eisenhauer et al., 1971). Evidently only about 30 to 35 days are needed by the adults to complete their flightless molt, and during this time the young grow rapidly, requiring about 50 to 60 days to attain flight (Mickelson, 1973). Goslings as heavy as three pounds have been observed killed by glaucous gulls, but most gull predation occurs on younger goslings.

Postbreeding behavior. During the summer molt, emperor geese congregate in large groups in favored localities. In midsummer there is a substantial molt migration of nonbreeders and probably also failed breeders where they concentrate on coastlines to complete their wing molts before undertaking the fall migration to winter quarters along the Aleutian Islands. "Herds" of up to 20,000 flightless birds thus have been seen on St. Lawrence Island during summer (Fay, 1961). Fay noted that about 5,000 birds were concentrated along one of the southern lagoons of St. Lawrence Island, out of a total summer population of 10,000 to 20,000 birds. In recent years an increasing number of molting birds have abandoned St. Lawrence Island for Siberia. A sizable portion of the Asiatic population of emperor geese also molt at the Ukouge lagoon on the Chukotski Peninsula, located on the northeastern coast of Siberia (Kistchinski, 1971; Hupp, et al., 2007).

Molting of breeding adults begins about 2 to 3 weeks after the young are hatched. Their flightless period occurs between mid-June and early August among birds from coastal Alaska, or considerably earlier than that of breeding adults that have hatched their young in late June or July. Arrival of fall migrants at Izembek Bay might occur as early as mid-August; these early arrivals are presumably also nonbreeders.

Status. The two breeding components of the emperor goose consist of an Alaskan segment that in the 1970s numbered about 60,000 to 75,000 breeders and an equal number of nonbreeders during fall, and a much smaller but uncertain number of breeding birds in Siberia (Bellrose, 1976). During that time an estimated total of 140,000 to 160,000 emperor geese were present during spring in the Yukon-Kuskokwim Delta. During the 1980s there was a sharp drop in US numbers, from 101,000 in 1982 to 42,000 in 1986 (Petersen, Schmutz, and Rockwell, 1994). By the early 2000s it was thought that no more than 80,000 might exist worldwide, although Rose and Scott (1997) made a world estimate of 165,000.

By 2009 the Alaska spring survey estimate or emperor geese was 91,900 birds, suggesting a 2 percent annual population increase since 2000 (Baldassarre, 2014). The fact that about 90 percent of the Alaskan population of emperors nest in the Yukon-Kuskokwim Delta, and thus well over half of the world's population occurs there, means that it is particularly important that the region (which also contains major concentrations of nesting spectacled eiders, brant, and cackling geese, among other waterfowl) be protected from disturbance, hunting, or uncontrolled development.

The US Fish and Wildlife Service's 2015 spring count of emperor geese was 98,200, 23 percent above the 2014 count of 79,800 and 49 percent above the 25-year average during 1981 to 2014. The spring index for emperor geese was the highest recorded in more than three decades (US Fish and Wildlife Service, 2015). Few emperor geese are shot by sport hunters in Canada or the United States, but subsistence hunting in Alaska accounts for several thousand birds annually, especially in the Yukon-Kuskokwim Delta (Wolfe and Paige, 1995; Wentworth and Wong. 2001).

Relationships. In a few respects, the emperor goose is one of the most aberrant of the "gray" geese (genus *Anser*), such as in its maritime foraging behavior and its unusual adult plumage pattern. Some genetic evidence supports the contention that it should be generically separated (*Philacte*) from the other typical species of *Anser* (Oates and Principato, 1994), but Livezey (1996) judged that the emperor goose is a sister species to the snow goose and Ross's goose and assigned it to the genus *Anser*, as I have done since the 1960s. The relatively long fledging period of the young (50–60 days) and their seeming poor resistance to extreme cold suggest that the species may have evolved from a much more temperate-adapted ancestor.

Suggested readings. Eisenhauer et al., 1971; Ogilvie, 1978; Owen, 1980; Petersen, Schmutz, and Rockwell, 1994; Kear, 2005; Baldassarre, 2014.

Greater white-fronted geese, adults and juvenile in flight

Greater White-fronted Goose

Anser albifrons (Scopoli) 1769

Other vernacular names. Laughing goose, specklebelly, tule goose, white-front

Range. Circumpolar; breeding from western and northern Alaska eastward across northern Canada to Hudson Bay, the western coast of Greenland, and in most of Arctic Eurasia except Scandinavia and Iceland.

North American subspecies. *A. a. frontalis* Baird: Pacific greater white-fronted goose. Breeds in western and northwestern Alaska from the Bering Sea coast east to Hudson Bay, wintering in California and adjacent western Mexico.

   A. a. gambeli Hartlaub: Tule greater white-fronted goose. Breeds in southwestern Alaska, probably from between Cook Inlet and the Alaska Range north through the Yukon-Kuskokwim Delta. Winters in the Sacramento Valley of California, sometimes with *frontalis*. *A. a elgasi* Delacour and Ripley is not recognized by the American Ornithologists' Union (1998) or by Kear (2005).

   A. a. flavirostris Dalgety and Scott: Greenland greater white-fronted goose. Breeds on the west coast of Greenland between 64° and 72°N. latitude. Winters mainly in Ireland and northwestern United Kingdom but occasionally reaches the northeastern United States, rarely farther south and west.

Measurements. *A. a. frontalis.* *Wing:* adult males 380–441 mm, adult females 362–419 mm. *Culmen:* adult males 44–56.5 mm, adult females 42–54 mm (Elgas, 1970); males 50.7–55.0 mm; females 47.8–50.9 mm (Ely and Dzubin, 1994).

   A. a. gambeli. *Wing:* adult males 441–480 mm, females 410–441 mm. *Culmen:* adult males 55–62 mm, adult females 49–59 mm (Elgas, 1970); males 58.3–59.7 mm, females 54.4–55.7 mm (Ely and Dzubin, 1994).

   A. a. flavirostris. *Wing:* males 410–455 mm, females 392–420 mm. *Culmen:* males 45–57 mm (Delacour, 1954).

Weights (mass). Pacific white-fronted goose: Nelson and Martin (1953): 22 males ave. 5.3 lb. (2,404 g), max, 7.3 lb. (3,322 g); 18 females ave. 4.9 lb. (2,222 g), max, 6.3 lb. (2,858 g). Ely and Dzubin (1994): 407 males, ave. 3,000 g.; 384 females, ave. 2,075 g.

   Tule white-fronted goose: Nelson and Martin (1953): 21 males ave. 6.6 lb. (2,993 g), max.7.5 lb. (3,402 g); 13 females ave. 5.6 lb. (2,539 g), max. 6.5 lb. (2,948 g). Swarth and Bryant (1917): 6 males ave. 7.25 lb. (3,288 g); 4 females ave. 6.31 lb. (2,861 g). Ely and Dzubin (1994): 54 males, ave. 3,000 g.; 53 females, ave. 2.700 g.

   Greenland white-fronted goose: Kear (2005): 223 winter adults, 1,800–2,900 g, ave. 2,409 g.

Map 4. Breeding (hatched, with denser concentrations inked) and wintering (shaded) distributions of the greater white-fronted goose in North America. The dashed line indicates the usual southern limits of breeding; stippling indicates peripheral breeding areas. In part after Baldassarre (2014).

Identification

In the hand. This generally brownish goose can be recognized by its yellowish to reddish bill, which lacks the black "grinning patch" of snow geese, and its yellow to orange feet. *Adults* of both sexes have a grayish brown head and neck, except for a white forehead and anterior cheeks, bordered narrowly with blackish coloration. The neck feathers are vertically furrowed, and the lower neck, back, breast, and sides are sooty brown, narrowly edged with white, producing a slight barring. The tail coverts and sides of the body behind the flanks are white, the tail is brownish black tipped with white, and the abdomen and sides are variably blotched with black. The scapulars and tertials are grayish brown, tipped with white; the upper wing-coverts are brownish gray (outwardly) or grayish brown and tipped with white, the white tips especially noticeable on the greater secondary coverts. The primaries and secondaries are bluish gray to black, while the underwing surface is slate gray. The bill is pinkish (orange-yellow in *flavirostris* of Greenland), grading to yellow (dorsally) and bluish (basally), and the legs and feet are orange. The tule race *gambeli* differs from the Pacific race in being darker, especially on the head and neck, with a longer neck, and a longer, deeper bill, but with less extensive dark underpart markings. The Greenland race also has dark brown upperparts but heavier spotting below than does the Pacific race. *Immatures* in their first winter lack the black ventral markings and white facial patch and have less conspicuous white markings on the flanks and upper surface. The abdomen is dull white with gray mottling, the bill is dull colored, and the legs and feet are pale yellow. The white forehead is gradually acquired the first winter, but heavy black underside markings are not attained before the second year, when the bill, legs, and feet are also still duller than in adults.

In the field. Both on land or water and in the air, white-fronts are notable for their rather uniformly brownish coloration, which is relieved by their white hindquarters and, at close range, by white foreheads on the adults. Sometimes their orange legs might be seen in flight, and usually at least a few of the birds in a flock will show black spotting underneath. The tule race is more prone to forage in marshes on tule (*Scirpus*) tubers than in upland fields, as is typical of the Pacific race. White-fronts are generally extremely wary birds and often utter a cackling *lee-leek* or *lee-lee-leek!* resembling taunting laughter while in flight.

Age and Sex Criteria

Sex determination. No plumage characters are available for external sex determination.

Age determination. Birds in their first year have little or no abdominal spotting and have yellowish feet and legs. Second-year birds are adult-like in plumage, but the color of the bill and legs is duller than in adults.

34

GREATER WHITE-FRONTED GOOSE

Distribution and Habitat

Breeding distribution and habitat. In Alaska the white-fronted goose breeds primarily in the northern regions and nests mainly near the coast. At Barrow and to the east it is a common coastal breeder, extending in marshy areas from one to twenty miles inland, with apparent centers of abundance at Smith Bay and the Colville Delta. White-fronts are also common nesters in the Kotzebue Sound region along the No-atak and Kobuk Rivers, and in the Yukon-Kuskokwim Delta region. The southern limit of breeding appears to be Cook Inlet. In Canada the species breeds from the Alaska boundary eastward to the Perry River, north at least as far as Victoria and King William Islands, and south to the Hanbury and Thelon Rivers.

The preferred breeding habitats are the muddy borders of small tundra lakes and the floodplains and mouths of Arctic streams, where broad flats often have grass-covered hummocks (Snyder, 1957). Dzu-bin et al. (1964) characterized the preferred nesting habitat as middle to low Arctic vegetation, in open tundra, the borders of shallow marshes and lakes, river banks and islands, deltas, dry knolls, and hillocks near rivers and ponds. Two major types of topography are used for breeding: coastal tundra with little surface relief and gently rolling upland tundra 50 to 700 feet above sea level, with lakes and ponds in the depressions. Willow- and shrub-fringed streams and ponds are used by white-fronted geese to a greater extent than by other geese.

Wintering distribution and habitat. In the United States, most wintering habitat occurs in the Central Valley of California and on the Gulf coast of Louisiana and Texas. In Mexico considerable numbers of white-fronted geese occur in northern and central areas, with a few as far south as the coasts of Tabasco and Chiapas (Leopold, 1959). There the birds prefer interior or coastal marshes or wet meadows and usually fly out to stubble to feed on fallen grain or green plant material. Alkaline flats and sandbars are not used as much as by snow geese.

In California, plains, fields, and swampy lowlands are used by *frontalis* for roosting, while foraging is done in open fields. However, the tule white-fronted goose (*gambeli*) reportedly inhabits marshes overgrown with tules (*Scirpus*), cattails (*Typha*), or willow (*Salix*), and rarely forages in grain fields. In these marshes the birds apparently feed primarily on the tubers and rhizomes of *Scirpus*, which they pull up from the bottom in water up to one and a half feet deep (Longhurst, 1955). The race winters in central California's Suisun marshes, in the Butte Creek Basin near Marysville, and on the Sacramento National Wildlife Refuge.

General Biology

Age at maturity. Ely and Dzubin (1994) summarized data indicating that tule white-fronted geese do not breed until their third summer, as was also reported for European birds by Boyd (1954, 1962), although some reports of younger females nesting do exist. Of 66 Greenland white-fronts, the average

Pacific greater white-fronted goose, adult

age of pairing was 2.5 years, and the average age of successful breeding was 3.2 years, with 28.6 percent first breeding at two years, 35.7 percent at three, 28.6 percent at four, and 7.1 percent at five (Warren et al. 1992).

Pair-bond pattern. Pair-bonds are unusually strong and permanent in these birds, as they are in other true geese. Inasmuch as fall and winter flocks are obviously composed partly of family groups (Boyd, 1953; Miller and Dzubin, 1965), it seems clear that pair and family bonds are persistent in this species. Contacts with parents during winter have been observed in 69 percent of yearlings, 39 percent of two-year-olds, and 38 percent of offspring at least three years old. Likewise, sibling bonds are strong and might be lifelong (Ely, 1993).

Nest location. Nests are usually situated on flats or on a slight hummock, often bordering a lake or stream (Snyder, 1957). Dzubin et al. (1964) noted that nests are seldom far from water. Typically the nest is located on a slight incline or at the top of a hillock, so that visibility of the surrounding area is not restricted. Conover (1926) likewise noted that all three nests he found were on small hills.

Clutch size. Kessel et al. (1964) reported an average clutch of 4.3 eggs for 12 nests in the Hooper/Igiak Bay area of the Yukon-Kuskokwim Delta, with an observed range of 3 to 6 eggs. Calvin Lensink (pers. comm.) found that 301 clutches also from the Yukon-Kuskokwim Delta averaged 4.86 eggs, with annual minimal and maximal averages of 3.72 and 5.32. A sample of 721 nests from the same region averaged 4.6 eggs, with annual averages ranging from 3.7 to 5.7 eggs (Ely and Dzubin, 1994).

Incubation period. The incubation length is somewhat variable, with most estimates ranging from 22 to 28 days, but including means of 25.0 days for 24 nests on the Yukon-Kuskokwim Delta and Alaska's North Slope, 24.0 days for seven nests in the Northwest Territories (NWT), and 23 days for nests on the Anderson River Delta, NWT (Ely and Dzubin, 1994). Most estimates for the European race *A. a. albifrons* are for 27 to 28 days (Cramp and Simmons, 1977), but that of the Greenland white-fronted goose has been judged at only 22 to 23 days (Fencker, 1950). This duration is close to the 21- to 22-day period determined by Brandt (1943) for a nest in Alaska. He noted that seven eggs were deposited in one nest during a ten-day period.

Fledging period. On the Anderson River Delta, Barry (1967) reported a 45-day fledging period. Ely and Dzubin (1994) specified a similar 42- to 49-day fledging period on the Yukon-Kuskokwim Delta, while Mickelson (1973) reported a 55- to 65-day fledging period for the same region. Warren et al. (1992) reported 43 to 45 days for Greenland birds.

Nest and egg losses. Considerable data are available on nesting success among North American white-fronted geese. Ely and Dzubin (1994) reported considerable variation in nest success (35.9–90.9 percent) over 11 years, with a collective average of 68.7 percent. These variations were associated with varied predation levels and climatic differences. Hansen (1961) noted a nesting success of 89 percent (eight

Greater white-fronted goose, adult at rest

of nine nests) in one year. Calvin Lensink (pers. comm.) found that young broods during the late 1960s and early 1970s in the Yukon-Kuskokwim Delta collectively averaged 3.94 goslings for 79 broods, suggesting that hatching success might at times be fairly high. Clutch sizes decline with advancing laying date and tend to be smaller in years of late springs (Ely and Raveling, 1984).

Important predators of adults and juveniles include eagles, coyotes, foxes, wolves, and great horned owls. Gulls, jaegers, Arctic foxes, grizzly bears, and weasels take eggs and goslings (Ely and Dzubin, 1994). Major avian predators on nests are probably jaegers, while glaucous gulls consume considerable numbers of young goslings. Foxes, especially red foxes, also account for the loss of some nests and young, as do eagles and snowy owls (Dzubin et al., 1964, Barry, 1967).

Juvenile mortality. Most data on juvenile mortality are from the Greenland and European populations of white-fronted geese. Boyd (1958) estimated a first-year annual mortality of 46 percent after banding, and about 43 percent for second-year birds, compared to an adult mortality rate of 34 percent. Among European white-fronted geese wintering in England, Boyd (1959) determined that from 1947 to 1959 the mean juvenile size ranged from 2.7 to 3.6 lb., and the proportion of young birds in the migrant population varied from 14 to 46 percent. Boyd believed that these marked differences in the yearly proportions of young birds resulted from variations in the percentage of adults that successfully bred rather than from annual brood-size differences.

Among American white-fronts, Miller et al. (1968) estimated a first-year mortality rate of 44.1 percent for Saskatchewan-banded geese and that juveniles were 2.4 times more vulnerable to mortality than were adults. The percentage of immature in migrating populations ranged from 11 to 38 percent and averaged 23 percent between 1960 and 1966. Adult survival averaged 65 to 70 percent in birds from the Pacific Flyway (Tim and Dau, 1979; Ely and Dzubin, 1994).

Adult mortality. Tim and Dau (1979) judged that Pacific Flyway white-fronts had annual mean survival rates of 65 percent (for adult females) and 70 percent (for adult males). Alaska birds produced somewhat higher estimates of 77.2 and 83.7 percent annual survival during two different multiyear banding periods (R. Trost, cited by Ely and Dzubin, 1994). Miller et al. (1968) estimated an average annual adult survival rate of 68.7 percent for Saskatchewan-banded geese. This compares closely to Boyd's (1958) estimates of 66 percent survival for adult Greenland white-fronted geese, and 72 percent for adult European white-fronts.

General Ecology

Food and foraging. Records of foods taken during winter are rather limited. Martin et al. (1951) listed a variety of cultivated grain plants (wheat, rice, barley) as important foods. Native plants that are taken include the vegetative parts of various grasses such as panic grass (*Panicum*), saw grass (*Cladium*), wild millet (*Echinochloa*), and the rootstocks of cattail (*Typha*), as well as sedges and rootstocks (tubers) of bulrushes (*Scirpus*). Hanson et al. (1956) noted that of six adults collected on their breeding grounds at Perry River, four had eaten horsetail (*Equisetum*) stems and branches, two had eaten blades or stems of cotton grass (*Eriophorum*), and one had consumed horsetail rootstalks. Barry (1967) found that 12 adults collected between June and August on the Anderson River Delta had been eating sedges and horsetail (*Equisetum*).

Sociality, densities, territoriality. White-fronted geese are relatively nongregarious and rarely occur in large flocks except perhaps during fall migration. Shortly after arriving at their wintering grounds they spread out and become inconspicuous (Miller et al., 1968). Breeding densities are generally very low; in the 1960s the Pacific Flyway population of some 200,000 geese nested over an area of about 40,000 square miles in western Alaska, whereas the Central Flyway population of some 70,000 birds nested over 84,000 square miles of northern and eastern Alaska, and 35,000 square miles of Arctic Canada (Dzubin et al., 1964).

Although not colonial nesters, white-fronted geese sometimes cluster for nesting in favored locations, and Dzubin et al. reported that breeding densities in the best habitats of the Yukon-Kuskokwim Delta region averaged 6 to 7 birds per square mile. Bailey (1948) noted that near Barrow, Alaska, the birds often nested in small colonies with 13 to 20 pairs present within a quarter mile.

In large areas of the Canadian Arctic, the estimated density was only 1 bird per 3 to 16 square miles. Averages for aerial surveys made during a six-year period indicate that in the Mackenzie River Delta breeding populations averaged 0.4 geese per square mile, while in the upland and coastal tundra areas between the Mackenzie and Anderson Rivers the average densities for the period were 1.2 to 1.4 birds per square mile. This illustrates well the tendency of white-fronted geese to favor upland nesting habitats over lowlands (Dzubin et al., 1964).

Interspecific relationships. Little specific information on possible interspecific competition between white-fronted and other geese exists. During migration, white-fronts often mingle with and forage with Canada geese and seemingly consume much the same foods, but only rarely are they seen among flocks of snow geese. Nesting in the Hooper Bay area of the Yukon-Kuskokwim Delta occurs in about the same habitats as are used by emperor geese, but the white-fronted geese show a distinct preference for nesting on small hills, whereas emperor geese nest on flatlands and closer to water (Conover, 1926). After hatching, white-front families move to inland tundra ponds, whereas emperor and cackling goose families use rivers and tidal sloughs.

General activity patterns and movements. During migration, white-fronted geese follow a very similar daily routine to that of Canada geese and often forage with them. Miller and Dzubin (1965) noted that two feeding flights are typical, one occurring in early morning and the other in late afternoon. White-fronts tend to be more wary than either snow geese or Canada geese, and this might serve to keep the species somewhat separated.

Social and Sexual Behavior

Flocking behavior. Large flock sizes are not typical of white-fronted geese, except perhaps during fall congregation and migration. Large flocks of molting birds do evidently occur in the vicinity of the upper Selawik River in northwestern Alaska, where groups of 2,000 to 5,000 birds have been seen on two large lakes (US Fish and Wildlife Service, 1956). During the gatherings of birds in their fall staging areas in western Canada, peak populations of 25,000 to 50,000 birds have been found on shallow lakes (Miller and Dzubin, 1965). Shortly after reaching their wintering quarters, however, the birds tend to spread out into smaller groups and become quite inconspicuous.

In studying the behavior of wintering flocks in England, Boyd (1953) reported that the wintering flocks there often numbered several hundred birds, but as flock sizes increased, their unity of behavior decreased, with the larger flocks tending to break up into smaller units that acted independently. Likewise during spring migration the flock sizes of birds passing through the Platte River valley of Nebraska are generally not very large, usually no more than a few dozen individuals.

Pair-forming behavior. Little has been written on pair-forming behavior, but it apparently consists of the gradual development of individual associations during the second winter of life, supplemented and strengthened by repeated use of "triumph ceremonies" between the paired birds (Boyd, 1954), although these "ceremonies" might not occur until spring or even summer (Kear, 2005). This same pattern is typical of geese generally.

Greater white-fronted geese, adult collective threat

Copulatory behavior. Copulation is preceded by mutual head-dipping associated with considerable tail-cocking and exposure of the white under tail-coverts. After treading, both birds again cock their tails, lift their folded wings, and call, with necks vertically stretched (Johnsgard, 1965).

Nesting and brooding behavior. Mated birds typically arrive relatively early at the nesting grounds and in pairs (Bailey, 1948). Nesting is initiated shortly after the arrival at the breeding grounds, usually in the second half of May. A high degree of synchronization of nest initiation and egg laying is not as evident in white-fronted geese as in the snow, cackling, and Ross's geese.

The female constructs a nest that is usually lined with mosses, grasses, and finally down. The male does not normally approach the nest closely but remains several hundred yards away. In spite of the birds' large size and their tendency to nest in hilly situations, the nests are extremely difficult to locate. Unlike Canada geese, but in common with emperor geese, incubating females usually do not attempt to leave the nest and sneak away unobserved at the first sign of danger. Instead, they suddenly flush from the nest when approached too closely. Even when the location of the nest is known, the brown plumage of the female so closely matches the dead tundra vegetation that it is nearly impossible to see her until she flushes.

With the hatching of the brood, the male joins the family and, at least in the Hooper Bay area, the families then tend to move to inland tundra ponds, well separated from families of emperor and cackling geese (Conover, 1926). Unlike snow geese, families do not flock together, and, when frightened, the goslings typically scatter and dive into the thick cover (Barry, 1967).

Postbreeding behavior. Limited data are available on molt migrations in the white-fronted goose. Nonbreeders and failed breeders leave nesting areas and move to low-Arctic areas that are free of disturbances and in appropriate habitats, such as those with low-relief islands, channels, meadows, marshes, and meandering streams (Alexander et al, 1991; Hohman et al., 1992).

Status. Bellrose (1976) reported that winter inventories from 1955 to 1974 indicated a North American population of about 200,000 white-fronted geese, while postbreeding season surveys in Alaska and Canada reported about 250,000 and 55,000 birds, respectively. The tule goose's population was much smaller. S. R. Wilber (cited in Bellrose, 1976) estimated that about 250 tule geese were present in 1955–56 among 2,000 white-fronted geese wintering in the Suisun marshes of California, one of its few known wintering areas. The total known population in the 1970s probably then numbered only about 1,500 birds (R. Elgas, pers. comm.) By 1990 the estimated population had grown to 8,000 birds and was increasing (Orthmeyer et al., 1996).

The US Fish and Wildlife Service's predicted 2015 fall population index for the Pacific population of greater white-fronted geese was about 379,000 birds, 25 percent below the 2014 index of 637,000, but the long-term trend showed no significant change. The midcontinent population is estimated from fall surveys in Alberta and Saskatchewan. That population estimate for the fall of 2014 was 1,006,000, 29 percent higher than the previous (2012) count of 777,900, and above the three-year average of 892,000. There was no significant decade-long (2005–2014) population trend (US Fish and Wildlife Service, 2015), but the trend based on fall surveys in the Canadian prairies was upward from 1992 to 2014. Survey data from 1975 to 2011 indicate a four-fold increase in the Canadian population of greater white-fronted geese, with a recent estimate of 3.5 million adults (Canadian Wildlife Service Waterfowl Committee, 2013, 2014).

During the 2013 and 2014 regular hunting seasons. the total estimated hunting kills of white-fronted geese in the United States were 256,369 and 339,559 (Raftovich, Chandler, and Wilkins, 2015). The Canadian midcontinent kill estimates (mostly from Alberta and Saskatchewan) for 2012 and 2013 were 59,470 and 75,113, respectively. Some additional birds are probably killed in British Columbia (Canadian Wildlife Service Waterfowl Committee, 2014) and by aboriginal hunters.

Relationships. The greater white-fronted goose is the most widespread species of *Anser* and is obviously closely related to the bean goose (*Anser fabalis*), the graylag goose (*Anser anser*), and especially the lesser white-fronted goose (*Anser erythropus*). Ploeger (1968) has discussed the possible role of glaciation patterns in the evolutionary history of these populations.

Suggested readings. Barry, 1966; Mickelson, 1973; Ogilvie, 1978; Owen, 1980; Ely and Dzubin, 1994; Kear, 2005; Baldassarre, 2014.

Snow Goose

Chen caerulescens (Linnaeus) 1758

Other vernacular names. Blue goose, wavy (Canada), white brant, white goose

Range. The snow goose breeds in northeastern Siberia (Wrangel Island) and along the Arctic coast of North America from Alaska through Canada and Canada's Arctic islands east to northeastern Greenland. Winters in North America along the Pacific coast from Washington to California, along the Gulf coast, the Atlantic coast from New England south to North Carolina, and in the interior from Arizona and northern Mexico east to Alabama. Recently increasingly wintering variably far north along the Mississippi River valley to the Ohio and lower Missouri River valleys.

Subspecies. *A. c. caerulescens* (L.): Lesser snow goose. Breeds in North America along the Arctic coast of Alaska and Canada's Northwest Territories and Arctic islands (southwestern Banks and southeastern Victoria Islands east to the Queen Maud Gulf and Adelaide Peninsula). Also breeds on Southampton Island, the western Hudson Bay coast south to Cape Henrietta Maria, and locally on the eastern Hudson Bay coast, and on Prince of Wales, Somerset, and southwestern Baffin Islands. Winters in the Pacific coast states from western Washington to southern California and adjacent Mexico, the southern Great Plains (Arizona to Texas and adjacent Mexico), the Gulf coast east to Alabama, and north in the Mississippi-Missouri Valley to Missouri, Iowa, and Nebraska. Also breeds in northeastern Asia, on Wrangel Island, Russia, a group that winters in California.

A. c. atlanticus (Kennard): Greater snow goose. Breeds in northwestern Greenland and on northeastern and high-Arctic islands, including western Ellesmere, Axel Heiberg, Bathurst, Cornwallis, Devon, northern Baffin, Bylot, Somerset, and Prince of Wales. Winters along the middle Atlantic coast from Long Island south to North Carolina.

Measurements. Lesser snow goose: *Wing* (chord): 45 males 395–460 mm, ave. 430 mm; 43 females 387–450 mm, ave. 420 mm. *Culmen:* 45 males 51–62 mm, ave. 58 mm; 40 females 50–60 mm, ave. 56 mm (Palmer, 1976). *Wing* (flat): 47 adult males, 428–474 mm, ave. 449 mm; 36 adult females 438–461 mm, ave. 430.2 mm (Trauger et al., 1971).

Greater snow goose: *Wing* (chord): 20 males 430–485 mm, ave. 450 mm; 10 females 425–475 mm, ave. 445 mm (Palmer, 1976). *Culmen:* 20 adult males 59–73 mm; ave. 67 mm; 10 females 57–68 mm, ave. 62.5 mm (Palmer, 1976).

Lesser snow geese, adult blue and white morphs in flight

Weights (mass). Lesser snow goose: 467 adult males, ave. 6.05 lb. (2,744 g), 522 adult females ave. 5.55 lb. (2,517 g) (Cooch et al., 1960). Ave. of 81 males (Queen Maud Gulf) 2,558 g, 289 females, 2,539 g (Mowbray, Cooke, and Ganter 2,000). Max., males 6.8 lb. (3,084 g), females 6.3 lb. (2,857 g) (Nelson and Martin, 1953).

Greater snow goose: 21 males, ave. 7.3 lb. (3,310 g), max. 10.4 lb. (4,717 g); 13 females, ave. 6.2 lb. (2,812 g), max. 6.5 lb. (2,948 g) (Nelson and Martin, 1953). Ave. of 87 first-fall birds, males 6.2 lb. (2,812 g), 85 females, ave. 5.5 lb. (2,495 g). Older cohorts, 10 males, ave. 7.6. lb. (3,447 g); 18 females, ave. 6.8 lb. (3,084 g) (Palmer, 1976).

Identification

In the hand. Snow geese are likely to be confused in the hand only with Ross's geese and possibly with immature white-fronted geese. Any white goose with a wing length of at least 400 mm, a culmen of at least 50 mm, and a body mass of more than 2,000 g would almost certainly be a snow goose. Ross's geese are likely to have wing lengths of less than 400 mm, culmen lengths of less than 50 mm, and a body mass of less than 2,000 g. The culmen–wing length ratio of Ross's geese averages 9.4, and that of lesser snow geese 7.6. The presence of a black "grinning patch" on the bill and the absence of both a bluish tint and warty protuberances at the bill's base would indicate a snow goose. Hybrids are likely to have intermediate weights and measurements; their bills often exhibit pale blush tinting basally. Young blue-morph snow geese might possibly be confused with young white-fronted geese, but their lack of a black grinning patch will serve to distinguish young white-fronted geese easily. Domestic white geese and swans are sometimes confused by hunters with snow geese; these birds lack black wingtips and have no black grinning patch.

In the field. Both in the air and on the ground or water, snow geese are readily identified by the partially or extensively white plumage, contrasting with the dark primary feathers. Ross's geese are identical in plumage color but are obviously smaller; have relatively short, thick necks and stubby bills; and rarely if ever exhibit the yellow to brownish facial staining that lesser snow geese often acquire while grubbing in mud for submerged rootstalks. Wild snow geese call almost constantly, and their rather shrill, repeated "*la-uk!*" notes are reminiscent of barking dogs, whereas the calls of Ross's geese are higher in pitch and are uttered more rapidly. Snow geese usually travel in larger flocks than do greater white-fronted geese, and even at a considerable distance the under wing-coverts of white-fronts appear nearly as dark as their primaries, while in blue-morph snow geese the anterior under wing-coverts are much lighter, and they also have much more white around the head. In flight the emperor goose might be confused with a blue-morph snow goose, but this morph does not occur within the range of the emperor goose.

Age and Sex Criteria

Sex determination. No plumage characters are available for sex determination, but external cloacal examination should suffice (see Canada goose account).

Map 5. Breeding (inked) and wintering (shaded) distributions of the snow goose in North America. The lesser snow goose's breeding range is shown below the dashed line; the greater snow goose's range is above. In part after Baldassarre (2014).

Age determination. The presence of a dull-colored, usually dusky bill, and legs and feet that are brownish to dusky, is indicative of a first-year blue-morph bird. Juvenile blue-morph birds also lack the elongated and contrastingly black-and-white patterned greater secondary coverts. Juvenile blue-morph birds have little or no white on the head, and their secondaries are dark brown, whereas white-morph birds are generally uniformly pale grayish in most body plumage and have mostly pale gray secondaries. In captivity, snow geese normally breed at three years of age, but sometimes they do breed in their second year of life (Ferguson, 1966). Thus, an open oviduct or a fully developed penis would indicate a bird at least two years old.

Distribution and Habitat

Breeding distribution and habitat. In Alaska, the breeding evidence for the snow goose is limited to a few, mostly old, records primarily from the vicinity of Barrow. Gabrielson and Lincoln (1959) also mentioned finding two nests near the mouth of the Kinak River in 1953. More recently a few small and scattered colonies have been found within 4 to 6 kilometers of the coast. In Canada, however, the nesting range is extensive, from the Mackenzie River delta in the west to northern Ellesmere Island in the north, Baffin Island and northwestern Greenland in the east, and Cape Henrietta Maria in the south. Within this range, the greater snow goose has the most northerly breeding distribution, including northern Baffin Island, Devon Island, Ellesmere Island, and adjacent Greenland (Snyder, 1967). Parmelee and MacDonald (1960) found the greater snow goose common on the Forsheim Peninsula of Ellesmere Island and reported that it was known to nest on Bylot, Devon, Somerset, and Axel Heiberg Islands as well as northwestern Baffin Island and Thule, Greenland.

The "blue" plumage morph of the lesser snow goose is slightly different genetically from the white morph, and probably resulted from geographic isolation during Pleistocene times. Climate changes after glacial melting allowed geographic dispersion, and eventual secondary contact of the two morphs occurred (Cooke, Parkin, and Rockwell, 1988). In Canada the blue morph nests from northern Hudson Bay to southwestern Baffin Island and north to Victoria Island (Parmelee et al., 1967). It has been observed in a few Alaska locations, such as along the Beaufort Sea, on the Seward Peninsula (Kessel, 1988), and on the Sagavanirktok River delta (a nesting report, Johnson and Troy, 1987).

Cooch (1963) reported that Bowman Bay, Baffin Island, had a 98 percent frequency of blue-morph birds in 1960, while the percentages were 82 at Cape Dominion and 53 at Koukdjuak, on southwestern Baffin Island. On Southampton Island the blue morphs composed 33 percent of all snow geese at Boas River, while at Eskimo Point on the west coast of Hudson Bay it was 15 percent. At Perry River, Northwest Territories, it was 12 percent, and blue-morph birds were present (1 percent) as far northwest as Banks Island. In the early 2000s, the percentages of blue morphs were little changed, with 76 percent on Baffin Island, 32 percent on Southampton Island, 25 percent on western Hudson Bay, 52 percent on southern Hudson Bay, and 75 percent on Akimiski Island (James Bay) (Kerbes et al., 2006). In recent decades a few blue-morph birds have been increasingly seen northwestwardly (Dzubin 1979), being reported as far west as Alaska's Seward Peninsula (Kessel, 1988), and have been seen nesting on the Sagavanirktok River Delta (Johnson and Troy, 1987).

Lesser snow goose, blue morph adult defending nest, Manitoba

The breeding habitat of lesser snow geese generally consists of low, grassy tundra associated with flat limestone basins or islands in braided deltas and is usually near salt water (Cooch, 1961, 1964). Snyder (1967) has characterized the breeding habitat as low, flat tundra, usually near lakes, ponds, or on river floodplains. The greater snow goose, however, typically nests in habitats where stony terrain meets wet and grassy tundra. On Bylot Island, the largest known colony of greater snow goose (about 157,000 birds in the 1990s) nests where the land is flat, marshy, and protected from the north by mountains (Lemieux, 1959; Reed, Giroux, and Gauthier, 1998).

Wintering distribution and habitat. Winter surveys performed by the US Fish and Wildlife Service from 1966 to 1969 indicate that of an average winter count of some 1.2 million snow geese, about 40 percent occurred on the Pacific Flyway and adjacent Mexico. About 25–30 percent occurred in the Central Flyway, with the same percentages present in the Mississippi Flyway; the remaining 5 percent (mostly greater snow geese) wintered on the Atlantic Flyway. In the Chesapeake Bay region, Stewart (1962) reported that the typical habitat of greater snow geese consisted of salt-marsh cordgrass (*Spartina alternifiora*), which fringes the coastal bays or occurs as islands within them and provides both food and cover for the geese.

The traditional wintering area of lesser snow geese in the Mississippi Flyway has been the coast of Louisiana. Their attraction to the mud flats along the Mississippi River delta has apparently been produced by the growth of various grasses and sedges (*Zizaniopsis, Scirpus, Spartina, Panicum,* and *Typha*) whose roots provide favored foods (Bent, 1925). Snow geese also commonly winter along the entire coast of Texas but mainly occur on the brackish marshes and low prairies. The greatest concentrations historically were in Chambers and Jefferson counties, where up to 300,000 or more birds sometimes occurred during the 1940s (Texas Game, Fish, and Oyster Commission, 1945). Sometimes considerable numbers also occurred in northern Mexico, along the coast of northern Tamaulipas, as well as in the interior *bolsones* (nondraining basins) of Chihuahua and Durango (Leopold, 1959). Some wintering in northern Mexico (south to Zacatecas and San Luis Potosi) still occurs.

The Pacific Flyway's wintering concentrations were traditionally centered in California, from the Tule Lake and Klamath areas in the north to the Salton Sea in the south, with major concentrations in the Central Valley. Since the 1960s the population has shifted northward, so that the majority of wintering birds now occur in coastal Washington and British Columbia. The Puget Sound region (Skagit River Delta) of Washington and the nearby Frazer River Delta of British Columbia are now important wintering areas for Pacific coast birds. This diverse range, from arid desert climates below sea level to moist and humid coastlines, encompasses an equally broad range of habitats. However, the common attraction would appear to be the availability of edible natural grasses or cultivated grain fields, with the bays, lakes, and marshes providing safe resting locations.

General Biology

Age at maturity. According to Cooch (1958), snow geese might first attempt to nest when two years old but succeed only under ideal conditions. Of 44 responses by aviculturists to a survey by Ferguson (1966), 11 snow geese began breeding their second year, 31 during their third year, and two in their

Lesser snow goose, adult blue morph in flight

fifth year. Lynch and Singleton (1964) concluded from age-ratio data that at least during favorable years the two-year-old segment of the adult flock must significantly contribute to the total breeding production. Barry (1967) observed that 17 percent of the geese he banded as goslings were on the Anderson River breeding grounds two years later. At the large La Pérouse Bay colony near Churchill, Manitoba, half of the successful breeders first bred when two years old, and by four years of age 86 percent were breeding (Cooke, Rockwell, and Lank, 1995).

Pair-bond pattern. Pair-bonds in snow geese are apparently strong and permanent, although remating is common after the death of a mate (Cooke, Rockwell, and Lank, 1995). Pairing between white- and blue-morph birds is common but not random, with the offspring of all types of mating equally viable (Cooch, 1961).

Nest location. Nesting of snow geese is typically in colonies, often numbering several thousand birds. Cooch (1964) reported nesting colonies exceeding 1,200 pairs per square mile and noted that the largest known colonies then were on Baffin Island, Banks Island, and north of Siberia on Wrangel Island. On Wrangel Island two nest location types are typical (Uspenski, 1966). One was the colonial type (averaging 12–64 nests per hectare), in which 114,200 nests occurred on 3,700 hectares (12 nests per acre). The other type consisted of small colonies or single pairs nesting among brant and Pacific common eiders near the nests of snowy owls (*Nyctea scandiaca*). In the case of the large colonies, the nests were protected by the concerted defense of the large number of birds, while in the second case the snowy owls, in protecting their own nests, also provided protection for the geese and ducks.

Soper (1942) reported that the nest is always placed on a slight grassy swell on the tundra, where the ground is relatively firm and well grown with mosses and grass. Most nests are built with plucked and shredded tundra moss and lined with fine grasses and down, while some are built with grass and chickweed and are smaller and less bulky than those made of moss.

Clutch size. Clutch sizes of both plumage morphs of lesser snow geese are the same, 4.42 eggs prior to any losses due to predation or other sources (Cooch, 1961). Uspenski (1966) indicated an average clutch of 3.27 eggs for 645 nests on Wrangel Island, with the highest clutch average (3.55) in areas of high nesting density, apparently reflecting relative predation losses. Eggs are laid in colonies over a 12-day period, and both morphs began and ended all their egg-laying within the same interval. However, white-morph birds tended to begin nesting slightly earlier than did blue-morphs, according to Cooch. Lemieux (1959) reported that 22 greater snow goose clutches averaged 4.8 eggs, with clutches of early nests averaging 2.5 eggs more than those begun only four days later. Attempted renesting has not been reported.

Incubation period. The incubation period averaged 23.6 days at the La Pérouse Bay colony (Cooke, Rockwell, and Lank, 1995). Cooch (1964) similarly reported an incubation period of 22 to 23 days for lesser snow geese. The female may lose 35 to 40 percent of her weight while incubating, compared to 15 to 20 percent for males (Kear, 2005).

Fledging period. Cooch (1964) reported that 42 days are required for obtaining flight in lesser snow geese, while Weller (1964) reported a 5.5- to 6-week fledging period, and Kear 6 to 7 weeks, with early-hatched broods growing faster and having higher recruitment rates than those hatched later. Lemieux (1959) likewise estimated a six-week fledging period for greater snow geese, while Lesage and Gauthier (1997) reported a 43-day period.

Nest and egg losses. Cooch (1961) noted that in an early (unusually mild spring) season an average of 19 percent of the eggs failed to hatch—from infertility, predation, flooding, or other causes. In a normal season, losses were 36.5 percent, and in a retarded breeding season 49 percent of the eggs. Major losses resulted from flooding, desertion, and dump nesting. Harvey (1971) also reported egg losses of 20 percent, mostly occurring late in incubation. Cooke et al. (1995) estimated that 30 to 60 percent of all eggs laid resulted in fledged goslings.

Juvenile mortality. Cooch (1961) reported that the average brood size at the time of hatching was 4.22 for 33 broods he studied in 1952. By the twelfth week the average size of the brood had been reduced to 3.33 goslings for 32 broods, or an approximate 12-week mortality of more than 20 percent. Lynch and Singleton (1964) presented productivity data on snow geese for the period 1949–59, indicating that winter samples reported average brood sizes ranging from 1.6 to 2.7 and percentages of immatures ranging from as low as 1.8 to 54.9 percent. The percentage of adult-plumaged birds accompanied by young varied from 1.6 to 75.7 percent, suggesting that in favorable years at least some two-year-old birds nest successfully. On the basis of such figures and banding studies, a probable 60 percent first-year mortality rate has been suggested (Cooch, 1963). Based on band returns, Rienecker (1965) estimated a first-year mortality rate of 49.1 percent.

Adult mortality. Cooch (1964) estimated that adult lesser snow geese have an annual mortality rate of about 30 percent, based on an analysis of banded birds. Boyd (1962) provided an independent calculation apparently based on these figures, and concluded that the lesser snow goose had an adult mortality rate of 27 percent, compared with a rate of 23 percent for the greater snow goose. This compares closely with a 22.5 to 25 percent adult mortality rate for the population of lesser snow geese wintering on the west coast (Rienecker, 1965).

General Ecology

Food and foraging. Foods of snow geese have been studied relatively little, and most available information is from the wintering areas. On the Atlantic coast, salt-marsh cordgrass (*Spartina alterniflora*) rootstocks were evidently major foods in the mid-1900s (Stewart, 1962; Martin et al., 1951). On the Gulf coast, a larger variety of foods were eaten, including the rootstocks of bulrushes (*Scirpus*), cattail (*Typha*), cordgrass, salt grass (*Distichlis*), the seeds and vegetative parts of square-stem spikerush (*Eleocharis quadrangulata*), and other herbaceous materials (Martin et al., 1951). Lynch (1968) noted that more recently the lesser snow geese of the Gulf coast had deserted the coastal marshes and their traditional foods to largely winter and forage in rice fields, cattle pastures, and other agricultural lands.

Glazener (1946) similarly noted that in the marsh areas of Texas, snow geese fed on reeds (*Phragmites*), salt grass, cordgrass, cattails, smartweed (*Polygonum*), and sedges (*Carex* and *Cyprus*) during the mid-1900s, while in prairie pastures they ate a variety of grasses (*Andropogon, Paspalum, Festuca, Eragrostis, Panicum, Setaria*, and *Sporobolus*). In the rice belt of Texas, snow geese sometimes consumed considerable amounts of rice.

This foraging behavior is not so true of greater snow geese, which still feed mainly on three-square (*Scirpus* spp.) sedge rhizomes (Bélanger and Béddard, 1994). Limited samples from the western states indicate that rootstocks of bulrushes, vegetative parts of cultivated wheat, and various other plants are taken. Several authors have commented that the birds' strong serrated bill is well adapted for pulling up and tearing roots.

Coues (cited in Bent, 1925) mentioned how the birds closely crop short grasses in the manner of domestic geese (*Anser anser*) and put to good use their tooth-like bill processes while pulling up and consuming roots and culms. Glazener (1946) also said that, unlike the Canada geese, which graze, snow geese are mainly "grubbers." Uspenski (1965) noted that while on their breeding grounds on Wrangel Island, the geese ate only the plants available in their immediate nesting area, and Barry (1967) reported a fairly catholic breeding-grounds diet, including sedges, ryegrass (*Elymus*), cotton grass (*Eriophorum*), willows (*Salix*), and horsetail (*Equisetum*).

Sociality, densities, territoriality. Snow geese are among the most social of all geese, and fall and winter flock sizes numbering in the tens of thousands of birds are not at all unusual. Cooch (1961) mentioned that strong female ties are present, at least through the first year. Such subadult birds remain with their parents until the latter's early stages of incubation, when they leave the breeding colony and molt on its periphery.

Densities of snow geese on their breeding grounds are sometimes almost incredible. Uspenski (1965) reported approximately 300,000 birds and 114,200 nests on Wrangel Island in 1960, which he believed represented the main world-nesting center for the species at that time. As noted earlier, these 114,200 nests occurred in an area of 3,700 hectares, or a nesting density of almost 8,000 per square mile. Cooch (1964) reported that he was aware of nesting concentrations of 1,200 pairs per square mile, allowing an average territory size of only about two acres per pair. Ryder (1967) noted nest densities of up to 4.61 nests per 1,000 square feet in preferred mixed (birch and rock moss) habitats of Arlone Lake in the Perry River area, but the average for mixed and birch-dominated habitats was about one nest per 1,000 square feet, the equivalent of 45 nests per acre.

Interspecific relationships. In general, snow geese form single-species nesting colonies, although Uspenski (1965) mentioned that on Wrangel Island the birds sometimes nest among brant, or close to the nests of snowy owls. Snow geese also sometimes nest in close proximity to cackling (previously "Canada") geese (Parmelee et al., 1967), and Nelson (1952) described at least two probable wild hybrids between these species. MacInnes (1962) remarked that the Richardson's cackling geese he studied at Eskimo Point that nested among the blue-morph snow geese suffered as many egg losses to jaegers as did those nesting outside the colony. Barry (1956) noted that, while the brant nested near the coastline of Southampton Island, the snow geese nested at least one-fourth mile behind the high tide line.

Major egg predators of snow geese appear to be the Arctic fox (*Alopex lagopus*) as well as jaegers, gulls, and ravens. Herring gulls (*Larus argentatus*) sometimes also are significant egg predators, as indicated by Manning's (1942) observations on Southampton Island and Harvey's (1971) later studies. Jaegers likewise are serious egg predators; Cooch (1961; 1964) mentioned that they destroyed all the eggs laid during the first two days of nesting at Eskimo Point in 1959, and also that they destroyed 49 percent of the eggs of brant and snow geese laid in poor habitat at Anderson River, Northwest Territories. Uspenski's (1965) clutch-size data suggest that egg predators are most effective in colonies with low densities or at the periphery of nesting colonies and provide a possible explanation for the colonial nesting tendencies of this species.

Snow geese in the Great Plains have not only increased tremendously in the past few decades but many have also shifted their spring migration pattern from the Missouri River to at least 100 miles west into the central Platte valley. This route change has brought more millions of snow and Ross's geese into contact with several million sandhill cranes, cackling geese, Canada geese, and greater white-fronted geese. Snow geese and other geese usually arrive in the valley slightly earlier than do the cranes, and are more prone to forage in the Rainwater Basin south of the Platte than to be concentrated like the cranes to the immediate Platte valley. However, snow geese certainly compete directly with both cranes and other geese for corn throughout that region, as well as with an expanding deer population.

Lesser snow geese, pair at nest with young, Manitoba

The geese and sandhill cranes have greatly benefited from the revolution in corn-growing technology in the Platte valley, where production increases of about six-fold followed World War II. Greatly expanded irrigation, fertilization, and chemical pest management have all combined to produce a corn-growing bonanza in that region, and helped make Nebraska one of the top corn-growing states in the country. By the end of the twentieth century, record-setting annual corn crops were being grown statewide, with the Platte valley responsible for nearly 40 percent of the state's total crop output. Between 1998 and 2003, the average annual state corn production was 1.1 billion bushels, and between 2004 and 2009, it averaged 1.4 billion bushels, or more than 150 bushels per acre. Assuming a harvesting efficiency of 90 percent, there would be about 15 bushels per acre left in the field for wildlife to consume.

This food bonanza modified the Central Flyway migration route of the snow goose, which historically had followed the Missouri River north through eastern Nebraska. As of 2016, several million snow geese were wintering annually from the lower Missouri–Mississippi valley west to northeastern Mexico, and migrating through the Central Flyway, with a major stopover in the Platte Valley. Estimates of maximum numbers staging during early spring in the Platte valley–Rainwater Basin region of east-central Nebraska have varied greatly but have ranged up to an amazing 7 million birds, with 1 million to 3 million commonly being estimated (Johnsgard, 2010, 2012a, 2012b), and as many as a million sometimes present on a single wetland.

Similarly, in the early 1970s an estimated 200,000 sandhill cranes were also staging in the Platte valley during spring. By the early 1980s crane numbers had increased to about 250,000, of which nearly all were Arctic-bound lessers. By the late 1990s, maximum spring crane counts in the Platte Valley had approached or exceeded 500,000 birds. Evidence has been accumulating that these mid-continent geese and sandhill cranes are now unable to accumulate the levels of fat reserves during their time in Nebraska that had been true in the 1960s and 1970s. This is probably the result of increased food competition among the millions of geese and cranes, and a progressively improving corn-harvesting technology. Like the snow geese using the Platte valley, the sandhill cranes are probably also now departing for their tundra breeding grounds in less than optimum breeding condition. But, like lesser snow geese, they have also benefited from the recently warming weather conditions in the Arctic during nesting and have continued to achieve successful reproduction.

General activity patterns and movements. Little specific information has been written on general activity patterns, which seem to be much like those of other geese. Roberts (1932) reported that in western Minnesota the spring migrants typically spent the night on a lake, left at sunrise, and fed until about 10:00 a.m. They then returned to the lake and waited until about 4:00 p.m. to come out once again to forage in stubble fields.

Glazener (1946) similarly noted that wintering snow geese in Texas typically left to feed in the morning somewhat later than the Canada geese, and most of them left *en masse*. They fed up to 30 miles from their roosting sites and moved to watering places in midmorning. Then they made a midafternoon flight to feed again and sometimes remained feeding until after dark. While the spring migration is typically a protracted one involving short daily movements and much local foraging activity, the fall migration across the continental interior is sometimes a nonstop flight to the wintering area. Cooch (1955) reported that in 1952 the population of lesser snow geese wintering on the Gulf coast flew nonstop from James Bay to Louisiana, an air distance of 1,700 miles, in less than 60 hours.

Social and Sexual Behavior

Flocking behavior. The large average size of snow goose flocks is well known; Spinner (1948) provided counts of a greater snow goose spring flock of 13,494 birds and a fall flock of 2,659 individuals. Musgrove and Musgrove (1947) noted that during the spring in Iowa, flocks of 15,000 to 20,000 are commonly seen in areas of concentration, while scattered flocks of 500 to 10,000 might be found between these concentration points. They gradually move up the river at the rate of about 20 miles a day, stopping at traditional concentration points that might at times hold up to a million birds.

Pair-forming behavior. Pairs are apparently formed in snow geese, as they are in other species of geese, by the increasing association of individual birds and the development of pair-bonds by the repeated performance of the triumph ceremony. This ceremony presumably occurs during the second winter of life, although the birds might not successfully nest until they are three years old. Pairing between color morphs is common but does not occur at random (Cooch, 1961; Cooke and McNally, 1975; Cooke, Finney, and Rockwell, 1976), favoring pairing with morph type of the parents, thus the incidence of intermediate-plumaged geese is relatively low. Sibley (1949) estimated that at least 10 percent of the migrant geese he observed in eastern Kansas consisted of intermediate birds. Cooch (1961) suggested that intermediate, (heterozygotic) individuals have been responsible for the northward spread by genes producing blue-morph birds, rather than through pioneering by entirely blue-morph birds.

During the 1950s, the incidence of blue-morph birds in the Southampton Island colonies was 30 to 35 percent. Over the 40-year period 1967–2006, blue-morph birds composed 27.5 percent of all the snow geese counted in the Great Plains region during Audubon Christmas Bird Counts. There is thus no evidence that either of these genetically based plumage types has shown a selective advantage over the other in correlation with changing Arctic climates. And, although Ross's geese historically were entirely of the white plumage phase, a few blue-morph adults have been found in recent years. This very rare plumage variant in Ross's geese, with a frequency estimated at no more than one in 10,000 birds (Johnsgard, 2014), probably has resulted from gene exchange during occasional hybridization with blue-morph snow geese.

Copulatory behavior. Copulation is preceded by the usual mutual head-dipping. After treading, the tail is not so strongly cocked nor are the wings raised so high as is typical of most species of *Anser* (Johnsgard, 1965).

Nesting and brooding behavior. The female constructs the nest with the materials at hand, usually mosses and grasses (Soper, 1942). Little down is present when the first egg is laid, but the down mat is luxuriant by the time the clutch is complete (Sutton, 1931). Only the female incubates, but the male stands close guard, often within 15 feet of the nest (Barry, 1956). The female rarely leaves the nest voluntarily during incubation but will forage some if driven off the nest (Manning, 1942). Manning reported that both sexes become very wary about four days prior to hatching, but after the young hatch the male becomes quite fearless. The female usually leads the young after hatching, while the male remains behind and protects the brood from intruders. Such families gather together into flocks containing about 40 adults and then leave the nesting grounds.

Postbreeding behavior. Adult birds undergo their molt while their offspring are still flightless, and during this time they might gather in fairly large flocks. Cooch (1957) described cases in which more than 15,000 flightless birds have been caught by being driven into large enclosures. Nonbreeding adults and subadults, having molted somewhat earlier than breeders, leave the breeding grounds about the time the young birds make their first flights, while adults and their young follow about three weeks later, or early September (Cooch, 1964).

Recent continental population trends. The two subspecies of snow geese differ appreciably in their numerical status, with the Atlantic coastal population of greater snow geese having once been threatened

Lesser snow goose, intermediate white-blue morph adult, Manitoba

with extinction but averaging about 800,000 birds during spring counts of the early 2000s. During the same period the midcontinent winter population of lesser snow geese averaged nearly 13 million adults, as compared with 2.3 million adults in the 1970s (Canadian Wildlife Service Waterfowl Committee, 2014).

Refuge management changes, altered agricultural practices resulting in far more grain production, and milder winters have all had major effects on midcontinental snow goose migration patterns, both as to timing, major wintering sites, and foods consumed. Far more snow geese now (2016) winter farther north, in the central and southern plains, than was the case in the 1970s. Many of these midcontinental birds are from the Central Arctic, Hudson Bay, and Southampton Island colonies.

For example, the Southampton Island snow goose population that I described in a 1975 book consisted of about 156,000 breeding birds in 1973. By 1979 it had grown to 233,000, by 1997 to 721,000, and by 2008 to 939,000. Other major Canadian lesser snow goose colonies include Baffin Island with 1.6 million breeding birds by 2005, the Canadian central Arctic with 1.45 million in 1998, southern Hudson Bay with 428,000 by 2007, western Hudson Bay with 246,000 by 2008, and the Canadian western Arctic with 420,000 by 2013 (Canadian Wildlife Service Waterfowl Committee, 2014).

In Canada's western Arctic colonies, about 400,000 snow geese were breeding on Banks Island (Egg River) and the Mackenzie River Delta and western Amundsen Gulf region, NWT (Anderson River and Kendall Island) in 2012, and wintering in the Central and Mississippi Flyways. In addition, from 60,000 to 140,000 lesser snow geese that nest on Siberia's Wrangel Island (70,000 in 2012) migrate through Alaska to winter in California's Central Valley, and increasingly on the Fraser Delta (British Columbia) and the Skagit Delta (Washington).

There were probably at least 12.5 million adult lesser snow geese alive in 2012, compared with 2.3 million during the 1970s (Canadian Wildlife Service Waterfowl Committee, 2014). Greater snow geese and Ross's geese have also expanded their populations. The high Arctic and more easterly oriented greater snow goose population that mostly breeds even north of the lesser snow goose and winters along the Atlantic coast first reached about 1 million birds by 2006. It had grown at an 8 percent annual rate since 1965 and attained an all-time high population estimate of 1.4 million by 2009.

Snow goose hunting. During regular hunting seasons of the late 1960s, American hunters were killing about a third of a million snow geese in the Mississippi and Central Flyways. At that time, the midcontinent population of snow geese totaled about 1.5 million birds. Since 1972, when both flyways had daily shooting limits of only four snow geese, the continental light goose populations have all shown almost continuous proportional increases, and hunting regulations have been relaxed accordingly. In the Central Flyway states between North Dakota and Texas, daily bag limits were increased from five to seven birds in the 1980s and to ten birds in 1992. The waterfowl-hunting season was also increased from 88 to 107 days, the maximum then permitted by the Migratory Bird Treaty between the United States and Canada.

Changes in this treaty during 1995 permitted the extension of the snow goose hunting season to March 10 in part of the Central Flyway. Other restrictions were also relaxed, including even larger or unlimited daily bag limits, allowing shooting to continue until a half hour after sunset, the use of electronic calls to help entice geese into shotgun range, and permitting more than three shells in shotgun magazines.

By the late 1990s the midcontinent snow goose population had reached about 3 million birds. It was eventually decided by conservation agencies of both the United States and Canada that the lesser and Ross's goose populations should be reduced to half the levels of the late 1990s, and the greater snow goose population to 500,000 birds. Annual kills of these "light" (snow and Ross's) geese in the United States and Canada gradually increased from an average of 581,000 during the 1980s to more than a million between 1998 and 2002, as hunting regulations were relaxed and seasons extended.

Special goose seasons during 1998–99 and 2001–02 thus helped increase the total continental light goose kill to an average of 1.3 million over the first four years of these "conservation" hunting seasons. During the 2013 and 2014 regular hunting seasons (excluding special regional seasons) the total estimated hunting kills in the United States were 305,575 and 396,634 snow/blue geese, respectively (Raftovich, Chandler, and Wilkins, 2015).

In spite of these control efforts, the total midcontinent population of snow geese grew from about 2.3 million birds in the 1970s to about 10.5 million birds between 2003 and 2012. In the early 2000s there were about 1.5 million adult lesser snow geese nesting in the central Arctic, about 250,000 in western Hudson Bay, about 500,000 in southern Hudson Bay, about 560,000 on Southampton Island, and about 1.6 million on Baffin Island. At the same time, the Ross's goose population had risen to 1.5 million to 2.5 million birds between the 1970s and 2014, nearly all nesting in the Queen Maud Gulf. Likewise, the greater snow goose population rose to about 80,000 to about 1 million birds, and the Wrangel Island population of lesser snow geese increased to about 150,000 birds (Canadian Wildlife Service Waterfowl Committee, 2014).

These progressively increased opportunities for late winter and spring goose hunting were balanced by an increasing wariness of the geese and other compensatory factors, since total light goose kills leveled off and have declined somewhat after reaching a high point of 1.55 million during the 1999–2000 season. For the four-year period 2010–13, the combined US and Canadian snow/Ross's goose kill averaged about 460,000 snow geese and 80,000 Ross's geese, or only about half of the record 1999 numbers (Canadian Wildlife Service Waterfowl Committee, 2014.).

Considering the continued spectacular increases in light goose numbers in spite of this high level of hunter-caused mortality, the targeted population levels just mentioned for all three goose populations are unlikely ever to be reached through even more liberalized hunting regulations and expanded seasons. It also seems unlikely that such changes that encourage ever more lenient goose killing have engendered any greater hunter understanding of, or respect for, their prey. In the spring of 2016, during the "conservation" spring hunting season on snow and Ross's geese, there were no daily or seasonal kill limits in Nebraska, and very few other hunting restrictions. In March, a pile of several hundred carcasses of snow geese (often disparagingly referred to as "sky carp" by local hunters), as well as three federally protected sandhill cranes, were found tossed into a roadside ditch with only their breast meat removed. So much for waterfowl hunters respecting their prey! The hunters were later identified and given token fines of up to about $250. It is not rare to see dozens of wounded snow geese present in meadows and cultivated fields along highways in the Platte River valley as late as May each spring.

Current populations. The US Fish and Wildlife Service recognizes three populations of lesser snow geese in their annual midwinter surveys, based on their breeding ranges (Wrangel Island, western Arctic, and midcontinent) plus one population based upon winter distribution (Western Central Flyway). Lesser snow geese and Ross's geese occur together in many wintering areas and are not differentiated during the Midwinter Survey (Canadian Wildlife Service Waterfowl Committee, 2014).

The Western Central Flyway population was estimated at 243,200 birds (excluding Mexico) in the 2015 midwinter survey and has increased 5 percent annually from 2006 to 2015. The collective western Arctic and Wrangel Island population is indexed by counts in California, Washington, and British Columbia. The 2014 winter population index for this region was 1,180,700 birds. This regional population complex has also increased an estimated 5 percent annually from 2005 to 2014.

In Canada's eastern Arctic, a Baffin Island population survey of nesting lesser snow geese was estimated to have about 1.6 million birds in 2005 along with 478,000 in the South Hudson Bay region. In 2008 more than 246,000 were found in the western Hudson Bay region, and 940,000 breeders were estimated on Southampton Island. Other surveys of Canada's Central Arctic colonies of the Queen Maud Gulf region indicated 1.46 million birds in 2006. By 2015 the lesser snow goose's overall population was estimated at 12 million to 14 million birds, and was then believed to be stable (Canadian Wildlife Service Waterfowl Committee, 2014, 2015).

The greater snow goose is monitored by spring counts in Quebec by the Canadian Wildlife Service. The 2015 count was of 818,000. This total was similar to the previous year and showed no significant trend over the previous ten years (US Fish and Wildlife Service, 2015). A Canadian estimate of the greater snow geese population in 2015 was that it was then fluctuating from about 700,000 to 1 million adults. In 2014 a spring hunting season on greater snow geese was instituted in several Atlantic Flyway states, with about 66,000 birds being killed. Similar spring seasons have been held in Canada since 1999, with about 20,000 to 50,000 birds being killed annually (Canadian Wildlife Service Waterfowl Committee, 2014, 2015).

Status. As of 2016, all the major snow goose breeding populations appeared to be at or near all-time highs, and most were seriously overabundant. However, in addition to global warming concerns, there are constant threats of oil and mineral development in the Arctic, dangers of disease among the great concentrations of birds on migration and in wintering areas, and the sensitivity of all Arctic-breeding geese to a series of disastrous breeding seasons when weather conditions are unfavorable, which require that close attention be paid to the management of these flocks.

Relationships. The snow goose and its very close relative the Ross's goose have often been placed (for example, by the American Ornithologists' Union) in a unique genus, *Chen*. This separate generic status is difficult to justify on any behavioral grounds and is not generally used elsewhere. Although there is no species of typical *Anser* that provides an obvious link-form with the snow geese, the snow and Ross's geese almost certainly evolved as a result of the isolation of various Arctic goose populations during late Pleistocene times.

Suggested reading. Cooke, Rockwell, and Lank, 1995; Mowbray, Cooke, and Ganter, 2000; Kear, 2005; Baldassarre, 2014.

Ross's Goose

Chen rossii **Cassin 1861**

Other vernacular names. None in general use.

Range. Breeds mainly in the Queen Maud Gulf Migratory Bird Sanctuary of the Northwest Territories, eastward along the gulf to at least 97°02′W latitude and southward in the interior to at least 66°21′N longitude (Ryder, 1967). Breeding also occurs on Banks and Southampton Islands and increasingly along the west coast of Hudson Bay south to James Bay. Wintered historically only in California, but recently has also been common throughout the Southwest and the southern parts of the Midwestern states, and less frequent along the mid-Atlantic coast, south occasionally to Florida.

Subspecies. None recognized.

Measurements. *Folded wing* (flat): Ave. of 31 breeding males 43.4 mm, range 39–49.3 mm. Ave. of 32 breeding females 368.5 mm, range 355–382 mm (Ryder and Alisauskas, 1999). *Wing* (flat): 47 adult males, 371–411 mm, ave. 394.8 mm; 32 adult females 362–391 mm, ave. 376.0 mm (Trauger et al., 1971). Range of 19 males (chord), 370–400 mm, ave. 385 mm; 20 females 360–395 mm, ave. 370 mm (Palmer, 1976).

Culmen: Ave. of 22 males 38.5, range 37–46 mm; females 37–40 mm (Delacour, 1954). Ave. of 47 breeding females 40.1 mm, range 35–45 mm (Ryder and Alisauskas, 1999). Range of 47 males, 37–46 mm, ave. 41 mm; 32 females 34–41 mm, ave. 38.5 mm (Trauger et al., 1971). Ave. of 52 breeding males 41 mm, range 39–49.3 mm, ave. of 20 females 38.5 mm, range 34–41 mm (Palmer, 1976).

Weights (mass). Ave. of 25 breeding males (early egg-laying period) 1,632.2 g, range 1,350–1,955 g; 20 early egg-laying females, ave. 1,640 g, range 1,151–2,040 g (Ryder and Alisauskas, 1999). Ave. of 18 males, 2.9 lb. (1,315 g), max. 3.6 lb. (1,632 g); 21 females, ave. 2.7 lb. (1,224 g), max. 3.4 lb. (1,542 g) (Nelson and Martin, 1953).

Identification

In the hand. Although the Ross's goose is normally found only within a limited winter and summer range, it occasionally strays far from its usual migratory route, and individual birds might turn up almost anywhere. If examined in the hand, Ross's geese exhibit a short bill (under 50 mm) that lacks an oval "grinning patch" and in adult males is usually warty near its base, which is bluish behind the

Ross's geese, adults in flight

Map 6. Breeding (inked) and wintering (shaded) distributions of the Ross's goose. In part after Baldassarre (2014). The species' historic Pacific coast wintering distribution is shaded; the recently acquired and increasing (as of 2016) winter expansions into eastern North America are indicated by dashed lines.

nostrils. Ross's geese also never exceed 4 pounds (2,000 g), and their chord wing measurements never reach 400 mm (415 mm if flattened). The mean culmen-wing ratio ranges from 9.6 to 10.2, compared with 7.6 in lesser snow geese. Among eight apparent lesser snow × Ross's geese hybrids (Trauger et al., 1971), the wings of four males averaged 415.8 mm, and four females averaged 393.5 mm. Their average culmen measurements were 45.1 mm and 46.6 mm, producing mean culmen-wing ratios of 9.2 for males and 8.5 for females. Blue-morph Ross's geese have been documented but are extremely rare (McLandress and McLandress, 1979; Johnsgard, 2014), perhaps as a result of gene exchange though hybridization or periodic mutations.

In the field. Ross's geese are best distinguished by direct size comparison with snow geese when they are in the same flock, or by their comparable size to large ducks, such as mallards. The bluish base of the bill is evident at fairly close range. Birds of intermediate size and appearance may fairly often be seen in wild flocks in the Great Plains, indicating that natural hybridization does occur and thus adds to the difficulties of field identification of Ross's geese among snow goose flocks.

Age and Sex Criteria

Sex determination. No plumage characters are available for external sex determination.

Age determination. Not yet closely studied but apparently comparable to the snow goose. In general, first-year birds are less conspicuously marked with gray than is the case with snow geese, and they are more difficult to recognize at comparable distances.

Distribution and Habitat

Breeding distribution and habitat. The initial discovery of the breeding range of the Ross's goose was in the Perry River area, and until the early 1950s the species was believed limited to that region. However, it was also found to breed on Banks Island (Manning et al., 1956), the McConnell River on the west side of Hudson Bay, and on the Boas River delta of Southampton Island (MacInnes and Cooch, 1963). Ryder (1967) later found many previously unknown colonies south and east of the Perry River and noted that they were all on islands in lakes. These rocks and shrubs of these islands provide protection from wind and also to some extent from rain and snow. Flat islands lacking such protection are avoided. Preferred lakes are not only sufficiently large to prevent predators from swimming across to the islands but also shallow enough (under 5 to 6 feet) to prevent ice bridges from being present at the start of the nesting season.

The primary nesting area of the Ross's goose is located within the Queen Maud Gulf Migratory Bird Sanctuary, Canada's largest federally owned and protected area, covering 23,848 square miles, which supports one of the largest nesting populations of geese (Ross's and lesser snow geese) on earth. The sanctuary is located about 80 miles south of Cambridge Bay, was established in 1961, and consists mostly Arctic tundra and marshes but includes some larger wetlands, such as Karrak Lake, the largest nesting colony of Ross's geese. The sanctuary supports more than 90 percent of the world's Ross's geese and 8 percent of Canada's lesser snow geese (100,000–700,000), as well as nesting or molting greater white-fronted geese (110,000), "Canada" (cackling?) geese (60,000), and brant (6,000). Ross's geese have rarely been reported west to northern Alaska along the Beaufort Sea, and west to the Sagavanirk-tok River Delta (Johnson and Troy, 1987).

Barry (1964) reported that Ross's geese nearly always nest on remote island-studded lakes, 8 to 40 miles inland, in fairly dry surrounding countryside. Less often they nest along rivers or on lakeshores. Ryder (1969) judged that the availability of food in the form of sedges and grasses is of major significance in determining the distribution of nesting colonies; also important are protection from flooding during the spring breakup and a source of nest cover in the form of shrubs or rocks. Islands that rise from 10 to 20 feet above water level but which have sufficient level places to allow for growth of food and nesting materials provide optimum nesting habitat.

Wintering distribution and habitat. The historic wintering location of Ross's geese is in central California, where they mix with and occupy similar habitats of the wintering lesser snow geese in the Sacramento and San Joaquin valleys, as well as nearer the coast in Ventura and Orange counties (Bent, 1925). Kozlik et al. (1959) noted that geese color-marked at Tule Lake wintered throughout the Central Valley but were not seen in the Imperial Valley, suggesting a possible different migratory route for birds wintering in that area. Marshall (1958) noted that following a mid-October arrival in the Klamath Basin, Ross's geese move to the northern San Joaquin Valley, where they remain until February or March.

The last few decades have resulted in a vast increase in winter Ross's goose records east of the Rocky Mountains as well as great increases in wintering snow geese (Prevett and MacInnes, 1972; Ankney, 1996; Drake and Alisauskas, 2004; Pearse et al., 2010). The earliest reports of the species in Nebraska occurred in the early 1960s, when a few began to appear in flocks of snow geese. By 1990 they could regularly be found among such flocks, and a tornado that occurred in York County in March of 1990 killed 1,200 white geese, among which 2 percent were Ross's geese (Johnsgard, 2013). In recent years as many as 5,600 Ross's geese have been observed in a single Nebraska location, but most are found scattered among snow goose flocks.

General Biology

Age at maturity. Ferguson (1966) reported that of eight respondents to a questionnaire, six reported initial breeding of Ross's geese at three years, and one each reported initial breeding in the first and second years of life.

Pair-bond pattern. Presumably pair-bonds are permanent in Ross's geese, as in snow geese. Ryder (1967) mentioned the strong attachment of males to incubating females and defense of the young; he also noted that yearlings retain family bonds until the incubation period of the next season's eggs is begun. Thus, it is evident that individual pairs must remain together throughout the nonbreeding period.

Nest location. Nests are built on various habitats and substrates, but Ryder (1967) established that preferred nest sites are mixed habitats of small birch stands and rocks, while pure rock or birch habitats

Ross's geese, two adults, with intermediate white-blue morph lesser snow geese, Missouri

have intermediate preference, and open habitats of low tundra have the lowest nest usage. Ryder concluded that sufficient protection from the elements and ample space for grazing determines nest density in a particular location. In the preferred mixed habitat types, nests had an average density of 9.5 per 1,000 square feet, with a maximum of 20.6 nests in this area, or only 50 square feet per nesting pair.

Clutch size. Ryder (1970b) reported a mean clutch size of 3.6 to 4.0 eggs prior to incubation in three years of study. Average clutch sizes in early nesting seasons averaged larger than those in late-starting seasons during these years. Nests started early in the nesting season averaged larger than those initiated only 3 to 4 days later. The interval between eggs averaged 1.5 days. Removing a few of the eggs from a nest did not seriously affect hatching of the remainder, but adding eggs to a completed clutch resulted in very low nesting success. Ryder (1970a) suggested that the small average clutch size of this species has evolved in relation to the food available to the female before arriving on the nesting grounds, as represented by the maximum increases in body weight that she can carry during her spring migration. A small clutch size thus avoids depleting the post-laying energy reserves of the female and correspondingly increases the probability of her efficient incubation and brooding of her eggs and young. Ryder found no evidence of attempted renesting.

Ross's goose, adult at rest

Incubation period. On the basis of 45 last eggs laid, Ryder (1967) determined the average incubation period to be 22 days, with a range of 19 to 25 days. No incubation occurs prior to the laying of the last egg, and only 2 percent of the nests had down present prior to the laying of the penultimate egg. After the laying of the last egg, however, 82 percent of the observed nests had down present.

Fledging period. Since freezing weather typically occurs between 40 to 45 days after the time of hatching, the fledging period is evidently slightly more than 40 days (Ryder, 1969) and has been estimated at about 45 days (Baldassarre, 2014).

Nest and egg losses. Ryder (1967) reported that of 351 eggs in 91 nests studied in 1963, 93.7 percent hatched, while in 1964 he determined a 79.2 percent hatch of 230 eggs in 59 nests. The percentage of eggs destroyed was remarkably low, averaging 2.2 and 14.4 percent for the two years, respectively, while the remainder of egg failures resulted from infertility or embryonic deaths. Arctic foxes caused high nest losses in 1964 at one locality, but avian predators caused few egg losses. In later studies, Ryder (1970b) reported yearly hatching success rates of 60.6 and 80.3 percent over two years.

Juvenile and adult mortality. Ryder (1967) noted that the average brood size of 99 broods from Perry River was 2.88 for broods not more than one week old. Fall flocks in Saskatchewan had an average of 2.72 young per family, and winter counts in California indicated an average of 1.65 young per family, suggesting a total decrease in brood size of 42 percent. Alisauskas and Rockwell (2001) estimated annual adult survival rates on more than 13,000 adults and 15,000 juveniles banded from 1961 to 1999 as 86.6 percent for adults and 54.0 percent for juveniles.

General Ecology

Food and foraging. Little has been written of the foods of Ross's geese. Hanson et al. (1956) reported that the gizzards of five birds collected on the breeding grounds included mostly sedges (*Eriophorum* and *Carex*) and some grass (*Poa*). Ryder (1967) examined 26 birds from the Perry River region and found some roots of grasses and sedges; leaves of grasses, sedges, and birch (*Betula*); and the stems and spikelets of grasses and sedges. Roots were consumed early in the season, while later on leaves and spikelets were utilized. No animal materials were found, even though several goslings were included in the sample. Dzubin (1965) noted that during the fall migrant geese in Alberta and Saskatchewan use large lakes for resting and fly out twice daily to wheat and barley fields, where they feed on waste grain.

Sociality, densities, territoriality. Sociality and associated densities on the breeding ground are even higher in the Ross's goose than in the snow goose. Dzubin (1965) noted that spring flocks are much smaller and more scattered than fall groupings moving through Saskatchewan and Alberta, with fall staging areas in the Kindersley district often reaching peaks of 10,000 to 20,000 birds in the early 1960s. Dzubin (1965) noted that in 1964 about 3,000 birds occurred on five small lakes, another 4,500 occurred on four lakes, and 1,700 were on three small saline sloughs. Temporary puddles 10 to 150 acres

Presumptive Ross's × snow goose hybrid, with blue-morph lesser snow goose, Missouri

in size and containing spike rush (*Eleocharis*) mats were used heavily for resting and feeding. Marshall (1938) mentioned a single flock of 8,000 Ross's geese in the San Joaquin Valley of California.

Breeding ground densities on preferred islands are often high; the total number of nests on five islands in Arlone Lake, Queen Maude Gulf Bird Sanctuary, was 769 in 1963 and 906 in 1964. These islands had an average density of 4.26 nests per 1,000 square feet. Observations of two pairs provided territory estimates indicating maximum territory sizes of 8 and 12 feet in open and rock habitats, respectively. Nesting begins somewhat earlier in higher than in lower concentrations (Ryder, 1970b). The breeding grounds concentration at 200-square-kilometer Karrak Lake in the Queen Maude Gulf Bird Sanctuary may represent the highest known goose density in the North American Arctic; in 2012 more than 750,000 Ross's geese and 450,000 lesser snow geese nested there (Alisauskas, Leafloor, and Kellett. 2012).

Interspecific relationships. Ryder (1967) investigated possible competition with snow geese for nesting sites on Arlone Island and concluded that both species avoid open situations and prefer edge areas of birch or mixed habitats. However, he could not find any definite evidence of competition, since

Ross's goose, adult preening

Ross's goose densities and clutch sizes were as high in regions of high snow goose densities as they were in areas where snow goose densities were low. Food was abundant, and interspecific aggressive interactions were uncommon. Ryder believed that a future substantial increase in snow geese could, however, alter nesting space for Ross's geese.

Ryder's studies indicated that, at least in his study area, avian nest predation was not a significant factor in affecting nesting success. However, Arctic foxes apparently not only sometimes kill adult birds but also might cause stress by harassment during egg laying, and sometimes cause great damage to nests. Ryder noted that 144 Ross's goose nests and 122 snow goose nests were destroyed in one week during 1964; this caused the desertion of one island-nesting colony.

In the wintering areas, Ross's geese initially mingle with snow geese and white-fronted geese but later tend to leave them and forage separately (Marshall, 1958). At this time they are associated mostly with cackling Canada geese, and feed mainly on green feed, whereas snow geese and white-fronted geese forage on rice fields and cereal croplands (Marshall, cited in Dzubin, 1965).

Hybrids between Ross's and snow geese have been described (Trauger, Dzubin, and Ryder, 1971; MacInnes, Misra, and Prevett, 1989; Kerbes, 1994), and apparent hybrids can often be observed in Central Flyway flocks of migrant geese (Johnsgard, 2014). A very few blue-morph Ross's geese have been reported in the Central Flyway (McLandress and McLandress, 1979; Johnsgard 2014).

General activity patterns and movements. Ross's geese are apparently very similar to snow geese in their daily activities and movements. Dzubin (1965) documented the gradual shifting of fall migration routes in western Canada to a more easterly direction, associated with a gradual loss of surface waters on the Canadian prairies since 1955.

Social and Sexual Behavior

Flocking behavior. Ryder (1967) noted that on their arrival at the breeding grounds, Ross's geese are in small flocks of 2 to 50 birds. These represent family groups or their multiples, and when incubation begins, the nonbreeders flock together, leaving the nesting grounds at the time of hatching to undertake their molt migration. Shortly after hatching, units of 2 to 15 families leave the nesting grounds and move to inland lakes and river courses. By three weeks after hatching such postnuptial flocks might number as many as 200 birds.

Pair-forming behavior. By the time they reach their nesting grounds, Ross's geese are apparently already mated, and no copulatory or courtship behavior was noted by Ryder (1967). Copulations have been observed during spring migration in April, although it is apparent that they could not account for the fertilization of eggs laid in June. Ryder commonly observed triumph ceremonies, and this behavior is known to be important in the formation and maintenance of pair-bonds in geese.

Copulatory behavior. The precopulatory behavior of Ross's geese consists of the usual mutual head-dipping of true geese, which is followed by treading. Postcopulatory posturing is relatively weak (Johnsgard, 1965).

Nesting and brooding behavior. Ryder's (1967) study indicated that nest building normally begins immediately after arrival at the nesting grounds and that considerable variation in nest construction occurs. During the egg-laying period the geese spend short periods at the nest site, with one bird grazing while the other defends the territory. The male usually leads the attacks, with the female immediately behind. During this time territorial disputes are at a maximum, but when incubation begins the colony becomes noticeably silent. Only the female incubates, while the male remains near the nest and defends it. Females incubate with the head held up, as in snow geese, rather than with the head and neck on the ground, as in the genus *Branta*. Unless disturbed, the female covers the eggs with down when leaving the nest. After hatching, the male defends the brood while the female leads them away from the source of danger.

Postbreeding behavior. As mentioned earlier, families rapidly merge into flock units, which may number several hundred geese within a few weeks after hatching. Loss of the flight feathers of adults is attained about 15 to 20 days after the peak of hatching in early July. Within three weeks of hatching, the young have sheathed tail and flight feathers emerging. By the end of August they are capable of flight, and the birds prepare to migrate south, usually among flocks of lesser snow geese. No significant molt-migration has been described in Ross's geese or snow geese.

Ross's goose, adult standing

Status. The historic population of Ross's geese might have been as few as 5,000 to 6,000 birds in the early 1900s. Between the late 1950s and the early 1970s, the estimated wintering population of Ross's geese was only about 100,000 birds. However, counts in the late 1960s on fall staging areas in Alberta and Saskatchewan, and on breeding grounds adjacent to the Queen Maud Gulf have indicated a population in excess of 300,000, suggesting that the number was then already increasing, an indication of the population explosion that both snow geese and Ross's geese exhibited over the next several decades.

During 1965–67, 37 mixed snow-Ross's goose colonies in the Queen Maud Gulf along the Arctic coast of Canada's Northwest Territories supported 44,300 nesting birds, with 77 percent of them Ross's geese. By 1988, there were 57 colonies totaling 467,000 snow/Ross's geese, with about 60 percent of them lesser snow geese. That 23-year period saw a 7.7 percent annual rate of population increase among Ross's geese and a 15.4 percent annual growth rate for lesser snow geese. Much of the snow goose's remarkable increase probably resulted from immigration out of colonies in the eastern Arctic, since a 15 percent rate of annual increase in geese is much higher than would be possible through local reproduction alone.

By 1998 about 495,000 Ross's geese were present in the Queen Maud Gulf Migratory Bird Sanctuary. The population of northwestern Canada reached or exceeded a million birds by 2001, and by 2006 the breeding population of Ross's geese in Canada's central Arctic was about 1.3 million birds. A recent Canadian estimate of Ross's geese in 2015 was that it fluctuates from about 700,000 to 1 million adults. The lesser snow goose's overall population was estimated in 2015 at 12.7 million birds and was designated in 1999 as overabundant (Canadian Wildlife Service Waterfowl Committee, 2014, 2015).

The Ross's population at Karrak Lake in the Queen Maud Gulf Migratory Bird Sanctuary was estimated at 659,600 in 2014, 22 percent higher than the 2013 estimate but not significantly different from the ten-year 2005–14 trend. There is also a large colony near the McConnell River, NWT (about 90,000 nesting pairs in 2005), and a small one (possibly up to 2,250 pairs) near Cape Henrietta Maria, Ontario. The total Ross's goose population (including immatures) was estimated at about 1.8 million birds in 2012 (Canadian Wildlife Service Waterfowl Committee, 2014). It was designated as overabundant in Canada in 2014 and subject to special control measures.

In recent years Ross's geese have composed at least 2 to 3 percent of the combined flocks migrating through Nebraska, judging from sample counts of migrant birds over the past decade. If at least 4 million snow geese were present in the Central Flyway in 2016 (see the snow goose account), Ross's geese might have then totaled at least 100,000 birds passing through Nebraska, and if there were then 700,000 to a million Ross's geese in North America, their number in the Central Flyway in recent years may have been as high as several hundred thousand.

In 2012 sport hunters in Canada killed about 35,000 Ross's geese during the regular 2012 hunting season (excluding those killed during the special white goose "conservation" seasons), and nearly 53,000 were killed in the United States. During the 2013 and 2014 regular seasons, the total estimated hunting kills of white geese in the United States were 52,769 and 91,823 Ross's geese, respectively (Raftovich, Chandler, and Wilkins, 2015). The total Ross's goose population was estimated at about 1.8 million birds in 2012, so the percentage of Ross's geese killed then by sport hunters represented only about 2 percent (Canadian Wildlife Service Waterfowl Committee, 2014). The 2015 midcontinent white goose population (snow plus Ross's geese) was estimated at 3,284,100, and winter population indices increased an average of 6 percent annually from 2006 to 2015 (US Fish and Wildlife Service, 2015).

However, as with snow geese, the great tendencies for mass flocking and the reliance on a few breeding and wintering areas for the vast majority of the species make the Ross's goose susceptible to rapid declines in population through disease, habitat disruption, and the like. Additionally, the increasing contacts and hybridization between Ross's geese and snow geese add another dimension of uncertainty to the future prospects of this species (McLandress and McLandress, 1979; Trauger et al., 1971).

Relationships. The Ross's goose is clearly closely related to the snow goose (Ryder and Alisauskas, 1995) and a fairly recent offshoot of an ancestral snow goose that probably differed little from the present-day form. Like the cackling goose, the evolution of a small body size was probably an adaptation to efficient breeding in areas with especially abbreviated breeding seasons, and perhaps also allowed for the exploitation of foraging opportunities open to a small-billed and small-bodied goose.

Suggested reading. Ryder, 1967, 1970; Ogilvie, 1978; Owen, 1980; Ryder and Alisauskas, 1995; Caswell, 2009; Baldassarre, 2014.

Canada Goose and Cackling Goose

Branta canadensis (Linnaeus) 1758

B. h. hutchinsii (Richardson) 1831

Other vernacular names. Cackling goose, Canadian goose, honker, Hutchinson's goose, Richardson's goose, tundra goose, white-cheeked goose

Range. Breeds across most of North America, from the Aleutian Islands across Alaska and northern Canada and south to the central United States. Resident or semi-migratory flocks of larger subspecies (*canadensis* and *maxima*) have also been established at many wildlife refuges and around parks and lakes in cities, often well beyond the probable original range of the subspecies. Introduced into Great Britain (where the 2015 population was about 200,000 birds), southern Sweden, and New Zealand (South Island). The "splitting" of *B. canadensis* into two species (Banks et al., 2004), with the small tundra-breeding forms now recognized as *B. hutchinsii*, has confused both the taxonomic nomenclature and the vernacular nomenclature but has been supported by genetic evidence (Baker, 1998; Scribner et al., 2003a, 2003b). The several geographically widespread populations of the latter group are now collectively called the cackling goose, with subspecies of both groups here assigned vernacular names to provide ease of reader recognition.

North American subspecies. (*Note:* Generally arranged from north to south and west to east breeding distributions.)

Canada goose
(Large-bodied group of Mowbray et al., 2002)

B. c. parvipes (Cassin): Lesser (Athabaska) Canada goose. Breeds in boreal forests and taiga from central Alaska east through the northern Yukon and Northwestern Territories to eastern Nunavut, and south to northern Manitoba, Saskatchewan, Alberta, and British Columbia. Winters mostly from northwestern Mexico through New Mexico, Texas, and Oklahoma north to Colorado and Nebraska (Mlodinow et al., 2008).

B. c. occidentalis (Baird): Dusky (western) Canada goose. Breeds along Prince William Sound, Cook Inlet, and inland through the Cooper River drainage, east to Bering Glacier. Winters south to western Washington and northwestern Oregon.

Canada geese, adults in flight

B. c. fulva Delacour: Queen Charlotte (Vancouver) Canada goose. Breeds along the coast and islands of British Columbia and southern Alaska, north to Glacier Bay; largely nonmigratory. Sometimes considered to be part of *occidentalis* Owen (1980).

B. c. interior Todd: Hudson Bay (interior) Canada goose. Breeds in northern Quebec, Ontario, and Manitoba around Hudson and James Bays, south to about 52°N latitude, and north to Churchill and the Hudson Strait. Has also colonized western Greenland as a molting site in recent years. Winters in the southeastern United States, from coastal New England south to South Carolina, and inland from Missouri east to New Jersey, and south to the Gulf coast.

B. c. canadensis (L.): Atlantic (common) Canada goose. Breeds in southeastern Baffin Island, eastern Labrador west probably to the east-west watershed, Newfoundland, Anticosti Island, and the Magdalen Islands. Winters variably far south through the eastern United States, mostly from New England to North Carolina.

B. c. moffitti Aldrich: Great Basin (Moffitt's) Canada goose. Breeds in the Great Basin of North America between the Rocky Mountains and the Pacific states, intergrading to the north with *parvipes* and to the east with *interior*. Also intergrades with *maxima*. Often considered a synonym of *maxima* (Palmer, 1976; Owen, 1980). Semi-sedentary, wintering varying distances southwardly, rarely to southern California and the Rio Grande valley.

B. c. maxima Delacour: Giant Canada goose. Historically bred on the Great Plains from the Dakotas south to Kansas, Minnesota south to Missouri, and western Kentucky, Tennessee, and northern Arkansas. Later (early 1900s) nearly extirpated but now widespread from the Great Plains eastward over most of eastern and central United States as a result of extensive restoration efforts. Semi-sedentary, wintering limited distances southwardly.

Cackling goose
(Tundra goose group of Mobray et al., 2002)

B. h. leucopareia (Brandt): Aleutian cackling goose. Until the 1970s, thought limited to breeding on Buldir Island (western Aleutian Islands), but later also found nesting on Chagulak Island (Islands of Four Mountains), and in the Semidi Islands (Byrd, 1998). Restored during the 1970s to Attu, Agattu, and Alaid-Niki islands (Near Islands). Winters in the Central Valley of California.

Map 7. Breeding and wintering distributions of Canada and cackling geese. The mapped limits for subspecies' ranges represent 1970s information. The indicated locations (dots) of B. c. moffitti *range represent 1970s introductions; it currently (2016) occupies the entire dashed area and somewhat beyond. The inked lines in northern Canada and Alaska indicate a recent and expanded estimate of* B. hutchinsii *southern breeding limits (after Mlodinow et al., 2008); stippling indicates extended northern breeding limits (after Canadian Wildlife Service Waterfowl Committee, 2014; 2015). Dotted areas in the southwestern United States, northern Mexico, and along the Gulf coast indicate midcontinental wintering areas of* B. h. hutchinsii; *wintering areas of other* B. hutchinsii *races along the Pacific coast are not distinguished from* B. canadensis. *Pointers indicate recent locations of the* B. h. leucopareia *population.*

leucopareia

minima

taverneri

Alaska

occidentalis   Yukon Territory   parvipes   N. W. Territories   hutchinsii
                                   (Mackenzie)   (Keewatin)

                                                                    Nfd.
                                                                    (Labrador)

fulva   Br. Col.   Alta.   Sask.   Manit.   interior   canadensis   Nfd.

Intergrades or
subspecies unknown

P.E.I.

N. B.   N. S.

Me.

Wash.   Mont.   N. Dak.   Minn.   Wis.   Mich.   Vt. N.   Mass.
                                                              H.
moffitti   N. Y.   R. I.
                                                                     Conn.
Oreg.   Ida.   S. Dak.   Penn.   N. J.

Wyo.   maxima   Ohio   Del.
        (presumptive original range and   W.   Md.
        1970s breeding localities)   Ill.   Ind.   Va.   Va.
Nev.   Utah   Colo.   Ky.
                        Kans.   Mo.   N. C.
Calif.   Tenn.   S. C.

Ariz.   N. Mex.   Okla.   Ark.   Ala.   Ga.

Baja   Miss.
        Son.   Tex.   La.   Fla.

        Chih.   Coah.

Baja
Sur   N. L.
        Sin.   Dur.   Tamps.
                Zac.
        Nay.   S. L. P.
              Gto. Qro.
        Jal.   Mex.   Pue.
        Col. Mich.
                Gro.   Oax.

B. h minima Ridgway: Ridgway's cackling goose. Breeds on tidal margins and coastal floodplains of the Yukon-Kuskokwim Delta in western Alaska. Winters mostly (since the 1990s) in the lower Columbia River Valley and Willamette Valley, Oregon.

B. h. hutchinsii (Richardson): Richardson's cackling goose. Breeds on Canada's Arctic islands from western Baffin Island west through Somerset, Prince of Wales and Victoria islands to Banks Island, and from northern Quebec west along the northern and western shores of Hudson Bay, including Southampton and Coats Islands, through Nunavut and Northwest Territories to the McKenzie River Delta. Winters mostly in the American Southwest, from southern Colorado to Durango, Mexico, and along the Gulf coast from Louisiana south to Tamaulipas.

B. h. taverneri Delacour: Taverner's (Alaskan) cackling goose. Breeding range uncertain but includes Alaska's Yukon-Kuskokwim Delta, the Seward Peninsula, and the northeastern Kotzebue Sound, and perhaps extends to the North Slope of Alaska and northwestern Canada, where the breeding birds might be *B. h. hutchinsii*. Winters mostly in the Willamette River Valley, Lower Columbia River Valley, and Columbia Basin.

Measurements. (Culmen and wing lengths)
(*Note:* Taxa are listed sequentially by increasing mean male culmen lengths. Average [mean] measurements are rounded to nearest mm.)

Canada goose

B. c. parvipes: *Culmen:* 70 males, ave. 42.4 mm, 59 females 40.6 mm (Mobray et al., 2002). *Wing* (chord): males (number unspecified) ave. 431 mm, females 410–423 mm (Aldrich, 1946). Ave. culmen-to-wing ratio (both sexes) 10.1.

B. c. occidentalis: *Culmen:* 130 males, ave. 46.3 mm, 131 females 43.5 mm (Mobray et al., 2002); *wing*: (sex unspecified) 395–450 mm (Delacour, 1954). Ave. culmen-to-wing ratio (both sexes) 9.4.

B. c. moffitti: *Culmen:* 18 males, ave. 49.6 mm, 3 females 45.7 mm (Mobray et al., 2002). Mean wing (number unspecified): males 517 mm, females 491 mm (Hanson, 1965). Ave. culmen-to-wing ratio (both sexes) 10.2.

B. c. fulva: *Culmen:* 175 males, ave. 51.2 mm, 34 females 47.6 mm (Mobray et al., 2002). Culmen 45–60 mm, wing 432–513 mm (Delacour, 1954). Ave. culmen-to-wing ratio (both sexes) 9.0.

B. c. interior: *Culmen:* 22 males, ave. 52.9 mm, 18 females 49.1 mm (Mobray et al., 2002). *Ave. wing* (chord): males (number unspecified) 491 mm, females 466 mm (Hanson, 1951a). Ave. culmen-to-wing ratio (both sexes) 9.4.

B. c. maxima: *Culmen:* 6 males, ave. 57.3 mm; 10 females, 52.6 mm (Mobray et al., 2002). *Ave. wing:* males (number unspecified), 540, females 500.5 mm (Hanson, 1965). Ave. culmen-to-wing ratio (both sexes) 10.2.

B. c. canadensis: *Culmen:* males (number unspecified), ave. 57.4 mm, females 53.4 mm (Mobray et al., 2002). *Wing* (chord): males (number unspecified), ave. 485 mm, females 465 mm (Aldrich 1946). Ave. culmen-to-wing ratio (both sexes) 8.8.

Giant Canada geese, pair protecting nest, Manitoba

Cackling goose

B. h minima: *Culmen:* 152 males, ave. 29.7 mm, 152 females 28.1 mm (Mobray et al., 2002). *Ave. wing (chord)*: both sexes 385 mm (Swarth, 1913). Ave. culmen-to-wing ratio (both sexes) 13.3.

B. h. leucopareia: *Culmen:* 36 males, ave. 36.6 mm; 46 females 35.1 mm (Mobray et al., 2002). *Wing* (sex unspecified): 358–405 mm (Delacour, 1954). Ave. culmen-to-wing ratio (both sexes) 10.6.

B. h. taverneri: *Culmen:* 60 males, ave. 37.8 mm, 61 females 36.1 mm (Mobray et al., 2002). *Wing* (sex unspecified): 365–424 mm (Delacour, 1954). Ave. culmen-to-wing ratio (both sexes) 11.0.

B. h. hutchinsii: *Culmen:* 129 males, ave. 39.0 mm, 125 females 37.7 mm; ave. wing: males 407 mm, females 389 mm (MacInnes, 1966). Ave. culmen-to-wing ratio (both sexes) 10.4.

Weights (mass).

(*Note:* Taxa are listed sequentially by increasing mean body mass.)

Ridgway's cackling goose: Ave. of 152 males, 1,546 g; 152 females 1,311 g (Mobray et al., 2002). Ave. of 30 males 4.4 lb. (1,005 g), max. 5.6 lb. (2,540 g). Ave. of 20 females 3 lb. (936 g), max. 5.1 lb. (2,313 g) (Nelson and Martin, 1953). Ave. of 24 autumn females, 1310 g (Kear, 2005).

Aleutian cackling goose: Ave. of 36 males, 1,945 g; 46 females 1,704 g (Mobray et al., 2002).

Richardson's cackling goose: Ave. of 129 males, 2,180 g; 125 females 1,920 g (Mobray et al., 2002). Ave. of 31 males 4.5 lb. (2,041 g), max. 6.0 lb. (2,722 g); ave. of 37 females 4.1 lb. (1,856 g), max. 5.2 lb. (2,359 g) (Nelson and Martin, 1953). Ave. of 25 autumn females, 1950 g (Kear, 2005).

Taverner's cackling goose: Ave. of 60 males, 2,606 g; 61 females 2,421 g (Mobray et al., 2002). Ave. of 4 males 4.95 lb. (2,241 g), max. 5.07 lb. (2,300 g); ave. of 5 adult females 4.54 lb. (2,059 g), max. 4.96 lb. (2,250 g) (Kessel and Cade, 1958).

Lesser (Athabaska) Canada goose: Ave. of 70 males, 3,266 g; 59 females 2,854 g (Mobray et al., 2002). Ave. of 184 adult males 6.1 lb. (2,766 g), max. 7.87 lb. (3,570 g); ave. of 194 adult females 5.45 lb. (2,471 g), max. 7.25 lb. (3,289 g) (Grieb, 1970). Ave. of 194 autumn females, 2,450 g (Kear, 2005).

Dusky (western) Canada goose: Ave. of 130 males, 3,232 g; 131 females 2,640 g (Mobray et al., 2002). Ave. of 36 adult males 8.28 lb. (3,754 g), max. 9.83 lb. (4,459 g); ave. of 26 adult females 6.9 lb. (3,131 g), max. 8.82 lb. (4,001 g) (Chapman, 1970). Ave. of 98 autumn females, 3,300 g (Kear, 2005).

Hudson Bay (interior) Canada goose: Ave. of 22 males, 4,472 g; 18 females 4,188 g (Mobray et al., 2002). Ave. of 44 adult males 9.28 lb. (4,212 g), max. 10.4 lb. (4,717 g); ave. of 45 adult females 8.3 lb. (3,856 g), max. 8.5 lb. (3,765 g) (Raveling, 1968b). Ave. of 36 autumn females, 3,330 g (Kear, 2005).

Queen Charlotte (Vancouver) Canada goose: Ave. of 175 males, 3,689 g; 134 females 3,043 g (Mobray et al., 2002). Ave. of 9 males 10.2 lb. (4,625 g), max. 13.8 lb. (6,260 g); ave. of 6 females 7.8 lb. (3,537 g), max. 9.5 lb. (4,309 g) (Nelson and Martin, 1953).

Atlantic (common) Canada goose: Ave. of 10 males, 4,921 g; 10 females 4,280 g (Mobray et al., 2002). Ave. of 232 males 8.4 lb. (3,809 g), max. 13.8 lb. (6,260 g); ave. of 159 females 7.3 lb. (3,310 g), max. 13.0 lb. (5,897 g) (Nelson and Martin, 1953). Ave. of 7 autumn females, 3,450 g (Kear, 2005).

Great Basin Canada goose: Ave. of 18 males, 4,017 g; 3 females 3,340 g (Mobray et al., 2002). Ave. of 10 adult males 9.9 lb. (4,334 g); ave. of 9 females 8.17 lb. (3,930 g) (Nelson and Martin, 1953). Ave. of 9 autumn females, 3,720 g (Kear, 2005).

Giant Canada goose: Ave. of 6 males, 4,858 g; 10 females 4.825 g (Mobray et al., 2002). Ave. of 13 captive-raised adult males 14.39 lb. (6,523 g), max. 16.5 lb. (7,483 g); ave. of 13 adult females 12.16 lb. (5,514 g), max.14.19 lb. (6,336 g) (Hanson, 1965). Ave. of 25 autumn females, 5,030 g (Kear, 2005).

Giant Canada goose with Richardson's cackling goose, Nebraska

Identification

In the hand. Even in juvenile plumage the distinctive dark head and neck with the lighter cheeks and throat are evident in both cackling and Canada geese. Combined wing and culmen measurements often separate the two, especially the culmen-to-wing-length ratio, which averages from 8.8 to 10.2 in *B. canadensis* and from 10.4 to 13.3 in *B. hutchinsii*. Additionally, the body mass of adult cackling geese rarely exceeds 5 pounds, whereas adult Canada geese rarely weigh less than 6 pounds.

In the field. Canada and cackling geese can usually be readily recognized by their black heads and necks, brownish body and wings, and white hind-part coloration. This combination also applies to brant geese, but those small geese are limited to coastal waters and can often be recognized by their short neck, dark bodies, and ducklike size. Cackling geese and the smallest races of Canada geese (for example, *parvipes*) also have relatively short bills and necks, with the neck length and bill length becoming progressively greater as the body size increases, so that the largest forms of Canada geese appear to be unusually long-necked and long-billed. The smaller taxa also have brief, high-pitched "yelping" calls, while the larger forms utter lower-pitched "honking" notes that often sound like *ah-onk*. However, vocalizations of the largest Canada geese include a wide variety of at least 13 adult call types (Whitford, 1998).

Distinguishing Canada geese from cackling geese in the field might be impossible at times, especially in the West, such as when visually separating the largest race of the Taverner's cackling goose (*taverneri*) from the smallest (lesser) race of Canada goose (*parvipes*) (Johnson, Timm, and Springer, 1979; Richter and Semo, 2006; Deviche and Moore, 2007; Mlodinow et al., 2008). These two taxa (*taverneri* and *parvipes*) possibly intergrade, making field identification even more difficult. Field marks that might help distinguish between them include the Taverner's shorter, stubbier bill; steeper forehead and more rounded crown profile; shorter, thicker neck; and somewhat smaller overall size (Deviche and Moore, 2007). Additionally, Taverner's cackling geese tend to be darker-breasted than lesser Canadas, are more likely to have a dark line along the midpoint of the throat ("gular stripe"), and exhibit more evident subterminal brown and terminal white banding at the tips of the upper wing-coverts (Mlodinow et al., 2008). In common with most Aleutian and some Ridgway's cackling geese, a narrow white collar is more often present at the base of the black neck in Taverner's cackling geese than in the lesser Canada (Mlodinow et al., 2008). These same general features sometimes also help distinguish several other races of the two species.

Age and Sex Criteria

Sex determination. Males average slightly heavier than females, but no consistent external plumage or soft-part differences appear to be present and usable for sex determination.

Age determination. *First-year* Canada geese can be recognized by one or more of the following criteria: notched tail feathers, an open bursa of Fabricius (a lymphoid sac opening at the roof of the cloaca that in large Canada geese averages 27 mm in depth (range 24–35 mm), a pinkish red area of skin around the vent, and, in males, a penis that is pink, less than 10 mm long, and neither coiled nor sheathed.

Hudson Bay Canada goose at nest, Manitoba

Second-year birds have tail feathers lacking notches, a bursa of Fabricius with an average length of 20.5 mm (range 18–24 mm), a pinkish red skin area around the vent, and, in males, a coiled and sheathed penis about 10 mm long and 4 mm in diameter when not extended. The tail feathers and flight feathers are shorter, narrower, and have less sheen than in adults. *Third-year* birds have tail feathers that lack notching, a bursa that is usually closed but might be open in about 40 percent of two-and-a-half-year-old geese, and a naked skin area around the vent that is flesh red to purple. *Adult* females in their third year or older have open oviducts (Hanson, 1949). Higgins and Schoonover (1969) reported that small Arctic-breeding geese can be aged with more than 90 percent accuracy by neck plumage characters. Adults of these geese have their black neck markings sharply demarcated from the pale breast, whereas in immatures the colors gradually merge.

Distribution and Habitat

Breeding distribution and habitat. Because of the extraordinarily great subspecific diversity in breeding habitats and the collective enormous breeding range of these races, no concise summary of distribution and habitat is possible for the Canada goose. Virtually all of the nonmountainous portions of continental Canada and Alaska might be considered breeding range, as well as the Great Basin of the United States, and recently the northern prairies as well. Reintroductions of Canada geese into refuges and other managed areas throughout the northern states have blurred subspecific distinctions and have confused the picture as to original versus current or acquired breeding ranges.

Canada geese (*B. c. canadensis* and *B. c. interior*) migrating in the Atlantic Flyway breed through an extensive area in eastern Canada. This breeding area consists of two major habitat types (Addy and Heyland, 1968): (1) the forest-muskeg zone of the James Bay lowlands (*interior*) and (2) the Arctic tundra on the upper Ungava Peninsula, Cape Henrietta Maria, the Belcher Islands, and other Hudson Bay islands. Birds wintering in the Mississippi Flyway breed throughout a large area of central Canada and are largely represented by the Hudson Bay race *B. c. interior.* Their breeding habitats are generally similar to those just mentioned for Atlantic Flyway birds and appear concentrated on the coastal strip of sedimentary deposits adjoining southern Hudson Bay (Hanson and Smith, 1950) but extend north to northern Manitoba.

Canada geese migrating in the Central Flyway consist of a complex of several breeding populations and subspecies. The larger forms include some Hudson Bay geese that breed to the west and southwest of Hudson Bay (Vaught and Kirsch, 1966) as well as some Great Basin Canada geese that breed on the prairies of western Canada and Montana. This population once included substantial numbers of giant Canada geese that bred in the tall prairies of the northern plains states. Restocking efforts have developed new populations of largely residential "giant" Canadas in areas from Kansas to the Dakotas and east to the eastern Great Lakes and Ohio River Valley, many of which have become suburban or even urban breeders.

Also migrating through the Central Flyway's "Tall Grass Prairie Population" (TGPP) are much smaller geese that include both the extremely small tundra-nesting Richardson's cackling goose *hutchinsii* and the slightly larger lesser (Athabaska) Canada goose, which also breeds southward through the boreal coniferous forests of northwestern Canada to northern British Columbia and east to northwestern Manitoba.

Somewhat to the west of this population, but also using the Central Flyway, is the "Short Grass Prairie Population" (SGPP) of small geese, which includes both the small races of both lesser Canada and Richardson's cackling geese that migrate through the high plains east of the Rockies. The breeding areas of the lesser Canada's population include a broad and diffuse area of the Northwest Territories. The eastern segment, the Richardson's cackling goose, breeds primarily along the Arctic Ocean coast at longitudes 100–110°W (Grieb, 1968). Birds breeding to the north along the coast of Victoria Island are also typical *hutchinsii* and are likewise tundra breeders. The western segment is composed predominantly of forest-breeding birds (presumably *parvipes*) that nest in the Mackenzie River drainage from 110°W longitude west to the Yukon Territory (and interior Alaska) and from about 58°N latitude north to the Arctic Ocean.

The remaining major contributor to the Central Flyway is the so-called "highline" population of Great Basin Canada geese (probably *B. c. moffitti*, or northern intergrades between that race and the Hudson Bay race *interior*). Typical Great Basin geese breed on the prairie areas of southwestern Saskatchewan, southern Alberta, and eastern Montana, while birds of uncertain racial status breed from about Portage la Prairie, Manitoba, westward to eastern Alberta and northward. Farther north, the lesser Canada goose breeds in boreal forest habitat across much of central Canada. As noted, the Richardson's

Lesser Canada geese, adults landing

cackling goose also migrates through the Central Flyway but breeds on coastal Arctic tundra along the west-central Canadian mainland (Nunavut and Northwest Territories), possibly south to the edge of the boreal forest (Grieb, 1966).

The Pacific Flyway likewise is made up of several population complexes. Along this flyway, migrations include at least six subspecies, of which the tiny Aleutian cackling goose probably has the longest route. The Ridgway's cackling goose and very similarly Taverner's cackling goose breeds on Bering coastal tundra along the western coast of Alaska. Away from the coastal tundra, and along such major rivers as the Yukon, Kuskokwim, Kobuk, and Colville, the lesser (Alaska) Canada goose is the typical breeding bird. Along the southern Alaska coast, around the Cook Inlet, Prince William Sound, and Copper River, it is replaced by a larger and darker race, the dusky (western) Canada goose, which breeds along this moist coastline southwest to about Bering Glacier (144°W longitude) with a maximum abundance in the Copper River Delta (Hansen, 1962).

Still farther south, along Alaska's coastal panhandle and on the adjoining islands and mainland of British Columbia, the even larger and more sedentary Queen Charlotte (Vancouver) Canada goose breeds in a wet-temperate climate and similar moist coniferous forest habitats. It is isolated by several hundred miles from the southern limits of *occidentalis* and nests from Ross's Sound near Glacier Bay south to northern Vancouver Island.

Finally, in the interior river valleys, reservoirs, and lakes of the Pacific Flyway states from the eastern slopes of the Cascades across the Rocky Mountains to Montana and south to California, Nevada, Utah, and Colorado, the Great Basin Canada goose breeds over a diffuse but extremely extensive area, intergrading eastwardly with the giant Canada goose of the Great Plains.

Wintering distribution and habitat. Wintering habitats vary less than breeding habitats, and it is not unusual to find representatives of three subspecies mixing on migration routes and on wintering areas. There is a general inverse relationship between the size of the bird and the distance between its breeding and wintering areas, with the smallest races (Aleutian, Baffin Island, Ridgway's, and Richardson's cackling geese) migrating to the most southerly wintering areas in California, while the largest forms (Queen Charlotte, giant, and Great Basin Canada geese) are often nearly nonmigratory and sometimes winter on their breeding ranges.

Definitions of typical wintering habitats no doubt differ according to region, but one useful analysis is that of Stewart (1962) based on studies at Chesapeake Bay. The habitat there is optimal because of the presence of extensive agricultural areas adjacent to open, shallow expanses of fresh, slightly brackish, or brackish estuarine bays, providing food in grain fields as well as in the shallow estuaries and providing roosting sites in the bays.

In the interior United States, the increasing numbers of large reservoirs that remain ice-free all winter and are adjacent to grain fields have resulted in an increasingly delayed fall goose migration and progressively more northerly wintering areas in recent years, at least for the larger subspecies. This combination of safe roosting sites and the availability of agricultural crops or other suitable foods would seem to be the prime requisites for wintering habitat. Reeves et al. (1968) documented such wintering population changes in the upper Mississippi valley for Illinois and Wisconsin geese.

Apparently at least part of the stimulus for the development of goose overwintering at Horicon National Wildlife Refuge in Wisconsin was the establishment of a resident flock and a reflooding of the marsh. A simple combination of food and sanctuary was responsible for developing the famous flocks of Canada geese at Horseshoe Lake Conservation Area in Illinois. Similar impoundments and refuge management programs have influenced the northern wintering limits of snow geese and Canada geese in the Great Plains, which increasingly winter variably north at least to Kansas, Nebraska, Colorado, and Wyoming, sometimes or locally to the Dakotas.

General Biology

Age at maturity. There might be individual or racial variation on the point of age at maturity. Two-year-old females of the larger subspecies no doubt occasionally breed; Craighead and Stockstad (1964) found that 27 to 36 percent of the wild female Great Basin Canada geese that they studied bred at this age, as did all three-year-olds. Brakhage (1965) indicated that a third of the two-year-old female giant Canada geese under observation nested, and Sherwood (1965) found that about three-fourths of such females produced eggs. Martin (1964) and Williams (1967) also reported breeding by two-year-old

Dusky Canada geese, pair protecting brood

Great Basin Canadas. Evidently nearly all two-year-old male giant Canada geese are capable of breeding, and a very small portion of yearling males might attempt to breed (Brakhage, 1965).

As to Arctic breeders, Williams (1967) also reported that some captive Aleutian cackling geese nested and reared young at that age. Even the small Canada geese breeding in the eastern Arctic (*B. hutchinsii*) often breed as two-year-olds, and a very few yearling males might also succeed (MacInnes and Dunn, 1988).

Pair-bond pattern. Canada geese are monogamous and exhibit strong pair and family bonds. Separation from a mate, or its death, will result in the formation of a new pair-bond, usually during the next breeding season (Hanson, 1965). Sherwood (1967) found that pair-bonds can be developed within a few hours in older, experienced, and "acquainted" geese, and these remained permanent as long as both remained alive. He found no polygamy, promiscuity, or pairing between broodmates. Pairing normally occurred on the nesting grounds, when the birds were two years old. Yearlings typically remained near their parents and rejoined them after the nesting season. Some yearlings formed temporary pairs, and broodmates retained their family bonds well into their second year.

Nest location. Nest locations vary greatly according to topography and vegetation. The same nest site might be used for several years (Martin, 1964). Hanson (1965) stressed the importance of muskrat houses as nest sites for marsh-nesting giant Canada geese, and in Manitoba common reed (*Phragmites*) is preferred over prairie grasses for nest construction (Klopman, 1958). Hardstem bulrush (*Scirpus acutus*) is a highly favored nesting cover in the western states (Williams, 1967).

Williams concluded that several factors contribute to favorable nest locations. These include good visibility, a firm and fairly dry nest foundation, a close proximity to water, adequate isolation, and nearness to suitable feeding grounds and brooding habitat. Dimmick (1968) noted that 72 percent of 45 Great Basin Canada goose nests he studied were on islands, these apparently being the nesting sites safest from predators. The highest nest density occurred near feeding areas, and 74.5 percent of the nests had excellent or good visibility. Sand was preferred over cobblestone for a nest substrate, and nests built over mud were elevated to keep the bottoms dry. The average distance to water was 45.7 feet, and shrubs or driftwood provided cover for the majority of the nests. MacInnes (1962) reported that tundra-nesting cackling geese also strongly favored small islands surrounded by open water, and with fairly hard, dry surfaces.

Clutch size. Among the larger races of Canada geese, the clutch size is fairly consistently centered around five eggs, with averages of various studies (Williams, 1967) ranging from 4.6 to 5.7. Weller (in Delacour, 1964) could find no correlation between clutch size and geographic location among nineteen studies of larger Canada geese. Fewer data are available on the Arctic-nesting cackling geese. MacInnes (1962) reported an average complete clutch size of 5.1 to 5.4 eggs for *hutchinsii*, and Gillham (cited in Spencer et al., 1951) reported an average clutch of 4.7 eggs for *minima*. Bellrose (1980) reported mean clutch sizes of 4.27 eggs for *minima*, 4.57 for *interior*, 5.22 for *maxima*, and 5.34 for *moffitti*. The rate of egg laying is slightly more than one day per egg in both the cackling goose (MacInnes, 1962) and the larger Canada geese (Williams, 1967).

Incubation period. Incubation periods may vary slightly geographically, with the largest and more southerly breeders possibly having slightly longer periods. The largest Canada geese require from 26 to 28.6 days (Hanson, 1965) for incubation, averaging 28 days (Williams, 1967). This compares closely with 28 days for *occidentalis* (R. G. Bromley, cited by Mowbray et al., 2002), and 25 to 28 days for *interior* (Kossack, 1950). Among cackling geese, a 26- to 27-day incubation period has been noted for *hutchinsii* (Jarvis and Bromley, 2000), as well as 24 to 25 days (MacInnes, 1962) and 24 to 31 days (Mickelson, 1975). Furthermore, the more southerly breeding races often attempt renesting (Atwater, 1959) if their first effort fails, whereas MacInnes (1962) found no indications of renesting in the Richardson's cackling goose.

Fledging period. Racial variations exist in fledging periods in relation to body size and length of the growing season. Hanson (1965) reviewed this relationship and noted that the giant Canada goose requires from 64 to 86 days (9 to 12 weeks) to attain flight. Other estimates include *interior* at 63 days (Hanson, 1997), *moffitti* at 7 to 8 weeks, and nine weeks for *interior* and *maxima* (Mowbray et al., 2002). In contrast, the cackling goose has a fledging period of only 6 to 7 weeks (Mickelson, 1975).

Nest and egg losses. Weller (in Delacour, 1964) summarized published data on nesting success in the larger Canada geese. The average of nine studies was a 67 percent hatch of total nests studied, with a

Canada goose, adult brooding young

range of 24 to 80 percent. Hanson (1965) reported a similar average nesting success rate of 58.6 percent, based on nine studies of birds he considered to be giant Canada goose, and a 71.1 percent average nesting success for eight studies of the Great Basin Canada goose. MacInnes (1962) reported a
high nesting success rate (75–90 percent) for *hutchinsii* during two years of study, although it is typical of Arctic-nesting waterfowl to exhibit great yearly fluctuations in productivity as an apparent result
of annual weather variations.

Juvenile mortality. Estimates of juvenile mortality based on brood size counts are not completely reliable, since brood mergers do occur. Hanson (1965) estimated an average brood size at the time of hatching, on the basis of all available data, as 4.2 young for the giant Canada goose and 4.1 for the Great Basin Canada goose. MacInnes (1962) reported that various studies indicated an 82 to 97 percent brood survival under wild conditions for the Great Basin Canada goose, and his Arctic studies on *hutchinsii* indicated a similar 85 to 90 percent brood survival during two years of study.

Following fledging, juvenile birds are subjected to considerably higher mortality than are adults, at least in part as a result of inexperience. MacInnes (1963) reported an annual mortality of 75 percent for *hutchinsii* juveniles, compared to 25 percent for adults in the tallgrass prairie flock. Likewise, Hansen (1962) estimated a 56.9 percent annual juvenile mortality rate for the dusky Canada (*occidentalis*), compared to a rate of 28.9 percent for adults. Martin (1964) noted a 47 to 64 percent mortality rate in first-year *moffitti*, compared to a 35 to 45 percent rate in adults, and Vaught and Kirsch (1966) estimated a 35 to 50 percent mortality rate of immatures (probably mostly *interior*) birds at Swan Lake, Missouri.

Adult mortality. Grieb (1970) has summarized reported mortality rates for various populations of Canada geese and calculated a 38.9 percent adult mortality rate for the shortgrass prairie population (mainly lesser Canada geese). Annual adult mortality estimates include lows of 25 percent among adults of the tallgrass prairie flock, and 25 to 30 percent among adult migrants at Swan Lake, Missouri, both of which consist predominantly of the Hudson Bay Canada goose (Vaught and Kirsch, 1966). Higher estimates of a 35 to 45 percent adult mortality rate have been made for the Great Basin Canada goose. The data of Martin (1964), Williams (1967), and Hansen (1962) also suggest annual adult mortality rates of about 30 to 40 percent for Canada geese in the western United States. Hunting intensities strongly affect mortality rates.

General Ecology

Food and foraging. Most studies of food habits of Canada geese are of wintering or migrating birds and might not be typical of breeders. Martin et al. (1951) summarized data from a variety of areas, indicating that the vegetative parts, particularly the rootstalks, of many marsh plants are consumed. Important plants include cordgrass (*Spartina*), salt grass (*Distichlis*), sago pondweed (*Potamogeton pectinatus*), widgeon grass (*Ruppia*), hardstem bulrush (*Scirpus acutus*), glasswort (*Salicornia*), and spikerush (*Eleocharis*). In a study of foods found in 263 gizzards and 31 crops from Lake Mattamuskeet, North Carolina, Yelverton and Quay (1959) found that sedges (mainly *Eleocharis* species and *Scirpus acutus*) made up 63 percent of the food volume, while grasses constituted nearly all the remainder, with common grains being most important. Likewise, Stewart (1962) found that waste corn was the food of primary importance for Chesapeake Bay geese wherever it was readily available, although sprout growth of various grain crops was also consumed, together with the vegetative parts of various submerged plants. In large estuarine bay marshes and coastal salt marshes, the stems and rootstalks of such emergent plants as three-square (*Scirpus americanus* and *S. olneyi*) and cordgrass are taken in large quantities.

Richardson's cackling geese, adults swimming, Nebraska

These relatively early studies fail to illustrate the contemporary significance of corn and other grain crops, which have become so important to migrating geese of several species, including Canada geese, snow geese, and white-fronted geese (Mobray et al., 2002; Baldassarre, 2014).

Sociality, densities, territoriality. Many recent studies, such as that of Raveling (1969a), have clearly established the fact that the basic social unit in Canada geese is the family. Raveling determined that adults and first-year young remained together all winter, and always reassembled if separated. When captured and released together, initial separation occurred, but in no more than 7.5 days the family was again intact. Rejoining of such families by yearling offspring of the past season was apparently fairly common.

Although such yearlings sometimes formed temporary pair-bonds during their second summer of life, these usually broke down and either the birds returned to their parents, or the yearling siblings remained together through the fall and winter. In some cases, permanent pairing occurred in late winter or early spring between birds that had formed temporary pair-bonds as yearlings.

With the assumption of a permanent pair-bond, the parental/family bond is finally broken, and the potential depends both on specific preferences on the part of both sexes and on relative male dominance in the vicinity of the female, as indicated by Collias and Jahn (1959). These authors believed

Ridgway's cackling goose, adult portrait

that sexual behavior such as copulation facilitated pair formation, and they also established that a bird could recognize the voice of its mate even when unable to see it. Pair and family bonds are maintained and strengthened by repeated use of the triumph ceremony (Raveling, 1969a).

Estimates of breeding densities are available from various areas and apparently vary greatly. MacInnes and Lieff (1968) found marked differences in nest density of *hutchinsii* in adjacent kilometer-square plots during the same year as well as considerable differences in density of the same plots in two consecutive years. The highest density they reported over the two years was 13 nests in a square-kilometer plot. Earlier (1962) MacInnes reported that the optimum breeding habitat at McConnell River supported up to 6 nests per square mile. In his 55-square-kilometer study area (21.2 square miles), he reported 129 nests in 1966 and 99 in 1967, or an average density of 4.7 nests per square mile.

Hansen (1962) reported some remarkable nesting densities of the dusky Canada goose in the Copper River Delta. In 1954 there was an overall average density of 6.4 successful nests (8.0 calculated for total nests) per square mile on an 88-square-mile area, while in 1959 one small (2.08 square miles) nesting area had an average density of 108 nests per square mile. This area of high-density nesting was limited to 12 square miles of river delta adjacent to the coast.

One of the finest goose nesting grounds in all of North America occurs over an 800-square-mile area of the Yukon-Kuskokwim Delta from Igiak Bay to about the southern tip of Nelson Island, where goose breeding populations (cackling, brant, emperor, and greater white-fronted) have averaged about 130 birds per square mile. In 1950 about 60 percent of these were cackling geese, or an estimated 78 birds per square mile (Spencer et al., 1951). In 1951 three study plots totaling 2 square miles in area had an average density of 153 nests per square mile, about 40 percent of which were those of cackling geese (Hansen, 1961).

Some examples of extreme nest-site proximity have also been reported for the larger and more southerly breeding forms, as summarized by Williams (1967). He noted a case of 11 western Canada goose nests placed on a single haystack in Oregon, and 31 nests on an island about 0.5 acre in size in California. Hansen (1965) also noted several other cases in which nest density ranged from 10 to 66 per acre. It would thus seem that basic territorial tendencies of Canada geese probably do not limit breeding densities or influence nesting distribution as much as do physical factors such as availability and distribution of suitable nesting sites.

Interspecific relationships. Little has been specifically noted as to relationships with other species and possible competition for food or nesting sites. In some areas cackling geese breed in close association with black brant (Spencer et al., 1951), while in some areas of eastern Canada they are found in association with snow geese. Some studies suggest that losses to predators of eggs and young are low compared to those resulting from flooding (Hansen, 1961), chilling, or other weather-related causes. Predators responsible for high nesting losses have included the coyote, red fox, Arctic fox, and striped skunk as well as birds such as ravens, crows, magpies, and various gulls and jaegers. Of these, probably only the mammals are effective predators after the goslings have left the nest, although jaegers are notable predators of small goslings.

General activity patterns and movements. Some studies on variations in activities according to time of day have been performed. Collias and Jahn (1959) noted that during the pre-egg stages, territorial activity is greatest early in the morning, as was also true of copulation frequency. All of the observed

Ridgway's cackling goose, adult walking

copulations were seen between 20 days prior to the laying of the first egg and the initiation of incubation. Canada geese typically fly out to forage in early morning and late afternoon in areas where they cannot forage at roosting sites.

Prior to taking flight, preflight intention movements, which consist of simultaneously lifting and shaking the head, are usually performed. Raveling (1969b) analyzed the occurrence of this behavior and found it tended to be given least and for the shortest time by single birds. The number of signals and the length of time from the first signal to takeoff were found to increase progressively for pairs and families of three and four birds, while families of five exhibited a countertrend. Raveling noted that whereas a gander did not always respond to head-tossing by members of his family, they always responded almost immediately to the gander. The importance of this signal in synchronizing and coordinating family activities is thus clearly apparent. Changes in vocalization speed and intensity, head-tossing or head-rolling movements, and the appearance of a conspicuous upper tail-covert pattern appear to be the major signals stimulating actual flight in nearly all other geese.

Social and Sexual Behavior

Flocking behavior. Probably the first suggestion of the importance of the family in the formation of larger flocks of geese was that of Phillips (1916), whose conclusions have been fully confirmed by later investigators such as Raveling (1969a) and Sherwood (1965). Raveling (1968a) compared flock substructure at the time of takeoff, while in steady flight, and at the time of landing, and concluded that only at the time of landing, when families almost invariably appeared together, did flock subunits clearly reflect actual family units.

Pair-forming behavior. As has been noted above, permanent pair formation typically occurs in two-year-old birds, probably in late winter or early spring. Mutual association of two birds and their coordinated performance of the triumph ceremony after aggressive encounters provide the basic means of establishing a pair-bond. Collias and Jahn (1959) described this process and noted that weather played a role in the intensity of pair-forming behavior, with cold weather tending to separate incipient pairs. After the selection of a nest site and associated establishment of a nesting territory, young of the past year are driven away from the parents, and the female and her nest site are defended from all intruders. The importance of male protection was illustrated by one pair in which the male died during the incubation period, and the female failed to hatch her young as a result of domination and disturbance from other pairs and unmated males.

Copulatory behavior. Copulation in Canada geese is preceded by mutual head-dipping movements resembling bathing. Copulation is usually initiated by the male, but the female soon participates and usually continues to neck-dip until the male prepares to mount her (Klopman, 1962; Johnsgard, 1965). Postcopulatory display is mutual and usually consists of raising the breast upward, extending the neck, and pointing the bill vertically upward while partially lifting the wings away from the body. Calls might be uttered by either or both birds.

Nesting and brooding behavior. Nest building is normally done almost exclusively by the female, although the male might very rarely participate to a limited extent (Collias and Jahn, 1959). In one instance noted by Collias and Jahn, a female built an entirely new nest from available materials in about 4 hours, and 45 minutes later had deposited her first egg. Down is usually added only after the first few eggs have been laid, and later some contour feathers might also be placed in the nest. To a limited extent nest-building behavior might continue throughout the incubation period, which helps prevent the nest from becoming flattened.

While incubating, the female usually leaves the nest only 2 to 3 times a day, to rest, forage, drink, bathe, and preen, and usually is gone for less than an hour at a time. The process of hatching requires about a day, and the young remain in the nest the first night. Females typically leave the nest with their brood the day after hatching but might bring them back to the nest for the next several nights for brooding (Collias and Jahn, 1959). Adoption of strange goslings is most likely to occur before they are a week old and if they and the parents' brood are of about the same age, after which the adults are likely to attack strange goslings. Fledging occurs at variable rates, correlated with adult body mass, and ranging from as few as 48 days in *hutchinsii* to 49–56 days in *occidentalis*, 63 days in *interior*, and 70 days in *moffitti*.

Postbreeding behavior. According to Hanson (1965), female giant Canada geese normally precede their mates in the postnuptial wing molt by 7 to 10 days, when the young are 30 to 50 days old. About 29 to 35 days are required for these large geese to regain flight, although 39 days were required for wing molting by a single adult male giant Canada goose studied by Hanson. Molting is most intense 35 to 40 days after the goslings have hatched (Baldassarre, 2014). Body molt continues long after the wing molt has been completed and might extend into spring.

Nonbreeding Canada geese often perform substantial migrations to areas where they undergo their molt, such as the barren grounds of the Thelon River Delta, Northwest Territories, where many large Canada geese might be seen in late summer (Sterling and Dzubin, 1967). Other populations undergo molt migrations as well. Many Canada geese molt on the west side of Hudson Bay (Davis et al., 1985), and some *interior* fly to molting grounds in western Greenland (Fox and Glahder, 2010). The dusky Canada goose might move to the western side of Cook Inlet, while the Queen Charlotte molts along Glacier Bay. The Alaska Canada goose perhaps molts along the Arctic coast of Alaska, the Athabaska Canada goose between the Mackenzie and Anderson Rivers in the Northwest Territories, and the Ungava Peninsula might be a molting area for Canada geese of the Atlantic Flyway (Sterling and Dzubin, 1967).

Status. The Canada/cackling goose population is probably much more secure at present than it has been in historical times. Winter surveys have suggested that the numbers have roughly doubled between the mid-1950s and the mid-1970s, and a fall population of about 3 million birds was estimated in 1974 (Bellrose, 1976). By 2002 the collective Canada/cackling goose's national population has reached 7.1 million birds (Moser and Caswell, 2004). It has been estimated that nearly 2 million giant Canada geese were present in the Mississippi Flyway by 2009, and in some regions population controls have been used to try to stabilize or reduce the population.

From 1999 to 2008, an average of about 2.57 million Canada/cackling geese were killed annually by hunters in the United States, and an average of 659,000 were taken in Canada (Baldassarre, 2014). During the 2013 and 2014 regular hunting seasons (excluding special regional seasons), the total estimated hunting kills in the United States averaged 2.58 million (Raftovich, Chandler, and Wilkins, 2015). Assuming a total North American population of about 7 million Canada/cackling geese, removing 2.6 million annually should eliminate about 35 percent of the population, or far more than the annual recruitment. Inasmuch as the national populations of these geese are not in evident decline, current estimates of the population and/or the hunting kill must need some adjustments.

In this regard, it has been reported that altering relative survival rates of adults has the greatest influence on population growth rates in giant Canada geese, followed by altering relative subadult survival rates, and lastly, altering relative nest success rates (Coluccy, Graber, and Drobney, 2004). This last influence may be more significant in the Arctic, where renesting efforts are unlikely because the short nesting season might prevent renesting.

The cackling goose's population was faring very well as of 2016, especially for the Aleutian race, which was once thought limited to about 300 birds on a single island (Buldir) where foxes, its major predators, were absent. During the 1970s some of these birds were transplanted to other Aleutian islands (Attu, Agattu, and Alaid-Niki) that were free of Arctic foxes. Still later, small native populations were found nesting on Chagulak Island and on two of the Semidi Islands. As a result of these and various

Ridgway's cackling goose, adult brooding

other intensive conservation efforts, by 2015 this race's population had increased to 189,100 birds, at an annual growth rate of about 7 percent per year since 2006. In contrast, the mainland (Ridgway race) of the Yukon-Kuskokwim Delta showed no significant population trend from 2006 to 2015. In 2015 it had an estimated population of 339,000 birds (US Fish and Wildlife Service, 2015). One extensive study of cackling goose survival rates estimated a 54.1 percent annual survival for adult males and 59.5 percent for adult females (Raveling et al., 1992).

The other populations of small Canada geese are also currently doing well. The 2015 breeding index for the lesser Canada goose based on limited surveys in coastal and south-central Alaska was 3,000, which was 77 percent higher than the 2014 index. An apparent downward ten-year trend (2006–2015) was not statistically significant (US Fish and Wildlife Service, 2015).

The Tallgrass Prairie Population (TGPP), probably mostly of Richardson cackling geese and some lesser Canada geese, averaged more than 431,000 birds from 2000 to 2009 and increased at an average rate of 5 percent annually (US Fish and Wildlife Service, 2009). The lesser Canada and Taverner's cackling geese have also been increasing at only slightly slower rates (3 percent annually) and numbered about 68,000 birds in 2009 (US Fish and Wildlife Service, 2009). Breeding ground surveys on Baffin Island from 1993 to 1999 indicated a collective breeding and nonbreeding population of about 100,000 cackling and lesser Canada geese (Dickson, 2000), the latter perhaps composed of nonbreeders or molting birds.

In 2014 the TGPP and Short Grass Prairie Population (SGPP) of Canada geese were administratively merged into the Central Flyway Arctic Nesting (CFAN) Canada geese because it is impossible to field-separate Richardson cackling geese from lesser Canada geese on their common wintering grounds. The CFAN group had an estimated 567,000 birds in the midwinter 2014 survey.

The population of Richardson cackling geese that nests on Baffin Island, the Hudson Bay coast, and eastern Queen Maud Gulf are now considered as an "East Tier" (formerly TGPP) population group of the CFAN, while those breeding in Victoria Island, Jenny Lind Island (northern Queen Maud Gulf), and the western Queen Maud Gulf are in the "West Tier" (formerly SGPP) population group. A major part of the East Tier population nests on the Great Plain of Koukdjuak in southwestern Baffin Island. Fall counts there from 1996 to 2009 ranged from about 124,000 to 202,000 birds, averaging about 160,000, and exhibited no long-term population trend during that period (Canadian Wildlife Service Waterfowl Committee, 2014).

The West Tier population group had an estimated spring population of 291,800 in 2015 and had increased an average of 9 percent annually since 2005 (US Fish and Wildlife Service, 2014, 2015). Likewise, midwinter counts of cackling geese in the Central and Mississippi Flyways from 1970 to 2012 showed an increasing trend from about 300,000 to 700,000, and overall estimates of midcontinent cackling geese indicate that from 1975 to 1979 there were believed to be about 368,000 adult birds, whereas from 2003 to 2012 about 3.3 million adults were believed present (Canadian Wildlife Service Waterfowl Committee, 2014).

Among the larger races of Canada geese, no significant population trend was found for the migratory North Atlantic Population (*canadensis*) of Newfoundland and Labrador breeders (130,000 birds estimated in 2014), while the similarly migratory Atlantic Population (*interior*) that breeds in northern Quebec, mainly along eastern Hudson Bay, Ungava Bay, and in the interior Ungava Peninsula) had a ten-year average of 187,000 breeding pairs in 2014 (Canadian Wildlife Service Waterfowl Committee, 2014).

The semi-resident large Canada geese of the Atlantic Population (*maxima*, *Canadensis*, and *interior*) had an estimated breeding population of 963,800 birds and had decreased an average of 5 percent annually during the prior decade (US Fish and Wildlife Service, 2015), perhaps because of encouraged intensive hunting pressures. The Southern James Bay Population of Hudson Bay Canada geese (*interior*) had a 2014 breeding population estimate of 81,000 birds and had declined slightly since 1990 (Canadian Wildlife Service Waterfowl Committee, 2014).

The Mississippi Valley Population of Hudson Bay Canada geese (*interior*) nesting in Ontario among the Hudson Bay lowlands west of Hudson and James Bays, had a 2015 breeding season estimate of 226,500 birds and did not show a significant decade-long trend. The Eastern Prairie Population of Canada geese (*interior*) in the Hudson Bay lowlands of Manitoba and Ontario had a spring 2014 breeding season estimate of 202,000 birds and a decreasing population trend since 1990. Boreal habitats in Alberta, Saskatchewan, Manitoba, and the Northwest Territories that were surveyed in the 1970s averaged about 156,000 Canada geese, compared with an average of about 550,000 from 2005 to 2014 (Canadian Wildlife Service Waterfowl Committee, 2014).

The Mississippi Flyway Population of introduced *maxima* made up the majority of Canada geese in this flyway, with 1.62 million birds estimated in 2015 and had apparently stabilized after many years of growth, possibly because of encouraged heavy hunting. The Western Prairie and Great Plains populations are also largely of *maxima* origins. Their combined estimated total spring populations in 2015 were 1.484 million, with an average annual increase of 9 percent since 2006 (US Fish and Wildlife Service, 2015).

The Hi-line Population of Great Basin Canada geese (*moffitti*) breeds east of the Rocky Mountains from southern Alberta south to northern Colorado. Breeding population surveys in 2015 indicated a population of 478,500 geese, a substantial increase from 2014 numbers but not statistically indicative of a long-term trend. The Rocky Mountain Population of Canada geese (*moffitti*) had an index survey of about 170,000 birds in 2015, with no apparent ten-year trend (US Fish and Wildlife Service, 2015).

The Pacific Population of Great Basin geese (*moffitti*) breeding west of the Rocky Mountains had an index survey of nearly 257,000 birds, with no significant decade-long trend. Recent Pacific Flyway breeding surveys of combined populations of lesser Canada and Taverner's cackling geese (*parvipes* and *taverneri*) have been geographically limited but suggest a possible long-term (2006–2015) decline. Earlier winter surveys from 2001 to 2007 have ranged from about 150,000 to nearly 700,000 birds, averaging about 430,000 (US Fish and Wildlife Service, 2014; 2015).

Considered a "Vulnerable" population in 2000, the dusky Canada goose (*occidentalis*) is surveyed mostly in the Copper River Delta of Alaska and has been doing notably well This representative component increased an average of 7 percent annually from 2005 to 2015. It had a breeding count estimate of 15,600 birds in 2014 (US Fish and Wildlife Service, 2014). The Vancouver Canada goose (*fulva*) has apparently not been systematically surveyed, but considering its similar range and ecology might have a comparable population.

In summary, the cackling goose's Arctic populations have generally thrived during the past decade, whereas the Canada goose's sub-Arctic populations (presumably mostly *parvipes*), have remained fairly stable. The larger and more southerly Southern James Bay and Mississippi Valley Canada goose populations (*interior*) have increased. The even larger temperate breeding populations (*moffitti* and *maxima*) are now so abundant that even with extended hunting seasons and other population management practices, they have steadily increased over the past three decades. During the decade 2004–13 the total annual estimated continental hunting kills of Canada/cackling geese have ranged from about 627,000 to 741,000 in Canada, and 2.1 million to 2.7 million in the United States, or roughly twice the numbers typical during the 1975–84 period (Canadian Wildlife Service Waterfowl Committee, 2015; US Fish and Wildlife Service, 2015).

Relationships. The cackling and Canada geese have been taxonomically separated by the AOU since 2004 (Banks et al., 2004). Genetic distinctions between these adjacent and broadly contacting (parapatric) breeding populations suggest that they became isolated about 1 million years ago (Shields and Wilson, 1987a; 1987b), or at about the middle of the Pleistocene epoch. The Hawaiian goose (*B. sandvicensis*) is also a near relative of the Canada goose, and these two species may likewise have been isolated for about a million years (Quinn et al., 1991).

Suggested readings. Hanson, 1965; Williams, 1967; Ogilvie, 1978; Owen, 1980; Mobray, Sedinger, and Trost, 2002; Baldassarre, 2014.

Barnacle Goose

Branta leucopsis (Bechstein) 1803

Other vernacular names. None in general use.

Range. Breeds in northeastern Greenland, Spitsbergen (Svalbard archipelago), and southern Novaya Zemlya. Winters in Ireland, Great Britain, and northern Europe, with infrequent winter occurrences in eastern North America.

Subspecies. None recognized.

Measurements. *Wing:* Both sexes 385–420 mm, ave. of males 410 mm, females 392 mm (Madge and Burn, 1988).
   Culmen: Both sexes 27–33 mm, ave. 29 mm (Madge and Burn, 1988).

Weights (mass). Twenty adult males in February, ave. 4.125 lb. (1,870 g), max. 4.625 lb. (2,098 g); 15 adult females, ave. 3.75 lb. (1,690 g), max. 4.5 lb. (2,041 g) (Boyd, 1964).

Identification

In the hand. Adults of both sexes have a white or buffy white head, except for a black stripe between the eye and bill, and a black crown which extends over the back of the neck and sides of the head to the base of the throat and continues downward to the back and lower breast, where it is sharply terminated. The back and scapulars are blackish and silvery gray, and the rump and tail are black. The tail coverts are white, and the white extends forward on the abdomen to the black breast and laterally on the sides to merge with pale barred gray flanks. The flight feathers are silvery gray, and the upper wing-coverts are gray with white tips and a subterminal black bar, forming a scalloped pattern. The bill, legs, and feet are all black.

In the field. Only an occasional visitor to North America, the barnacle goose nevertheless has appeared in a surprising number of localities, mainly along the eastern coast. It is slightly larger than a brant and differs from it in having a predominantly white head and a light gray upper-wing coloration rather than dark grayish brown. The underwing coloration is likewise light silvery gray and much lighter than that of the brant. The extension of the black neck color over the breast readily separates the barnacle goose

Barnacle geese, adults in flight

from the Canada goose, even at a great distance, and the contrast between the dark and light parts of the body is much greater as well. Its usual flight call is a barking, often repeated *gnuk*; a flock sounds something like a pack of small dogs.

Age and Sex Criteria

Sex determination. No plumage characters are usable for external sex determination.

Age determination. The presence of gray flecking on the head and a somewhat grayish rather than entirely black neck will serve to identify first-year birds. The black and white markings on the upper surface of the wings are also less well developed in first-year birds so that the upper-wing surface appears somewhat duller and darker. First-year birds are generally paler than adults and usually have more black on the head. A few brown feathers may be left on the wing coverts by the second summer, and thus these birds might be aged until August (Kear, 2005).

Natural History

Habitat and foods. Barnacle geese are in general confined as breeding birds to Arctic areas where cliffs or rocky slopes are located close to lakes, rivers, marshes, fjords (upper portions), or even coastlines, but they also sometimes nest on level ground, in the vicinity of brant or eiders. For the rest of the year the birds are essentially maritime, foraging on tidal flats, coastal marshes, and adjoining grassy areas. They are essentially exclusively vegetarians as adults and feed largely on grasses, sedges, and the leaves of various herbs and shrubs, including even the leaves, twigs, and catkins of Arctic willow (*Salix arctica*). Among 14 birds obtained during winter in England, over 90 percent of the food remains consisted of grasses, and no animal materials were present (Palmer, 1976).

Social behavior. It is probable, but not certain, that barnacle geese mature at the age of two years, when full adult size and plumage are reached. Records from captivity indicate that few birds nest as two-year-olds, and most breed when three (Ferguson, 1966). Pair-formation behavior in captivity indicates that the hostile behavior patterns and triumph ceremonies are nearly identical to those of Canada geese. The two species likewise perform preflight head movements that effectively expose their white cheek and throat areas. The precopulatory displays consist of mutual head-dipping movements, and although during the postcopulatory display the male raises his folded wings somewhat more strongly than do Canada geese, the posturing is otherwise almost identical (Johnsgard, 1965). It is assumed that most pair-forming behavior occurs on the wintering grounds, and the birds are well paired by the time of their arrival at the breeding areas. There is a high degree of long-term monogamy (Black, Choudhury, and Owen, 1996).

Barnacle goose, adult portrait

Barnacle geese, pair walking

Breeding biology. On Novaya Zemlya, these geese typically seek out rocky outcrops, ledges of steep cliffs, or the tops of low hills for nesting, which is done in scattered colonies that are highly conspicuous (Dementiev and Gladkov, 1967). On Spitsbergen (Svalbard) and in Sweden nesting is typically on islands to reduce risks of predation. Observations in northeastern Greenland by Ferns and Green (1975) indicate comparable nest sites, as on a nearly vertical basalt exposure about 50 meters high, where the birds chose flat and gently sloping ledges about 75 centimeters deep and 1 or 2 meters wide. Evidently nest sites are often used year after year, and gradually accumulate nesting materials. The clutch size typically ranges from 4 to 6 eggs, but up to 9 have been reported. The inaccessibility of nests to most ground predators probably reduces nesting losses and is presumably the reason for the cliffside nesting; even the geese have difficulty landing on the small ledges. Incubation is performed by the female alone and requires 24 to 25 days.

Barnacle goose, adult brooding young

There is evidently a rather high incidence of non-nesting or unsuccessful nesting, judging from the low percentages of goslings reported both on the breeding grounds and in wintering areas; Ferns and Green (1975) noted average brood sizes of 2.7 to 2.8 young for single families as well as for amalgamated family parties, but only 11.1 percent young were seen in the total population. They calculated that about 90 percent of the adult population either did not attempt to breed or failed in their efforts. On Svalbard nesting success varied from 18 to 74 percent in different years in one study, and breeding success tends to increase with age with the highest success among pairs 6 to 11 years old. There is a very low incidence (2 percent) of divorce among wild birds, and mate choice has a significant effect on breeding success (Black, Choudhury, and Owen, 1996). The fledging period, based on captive birds, is 6 to 7 weeks, a relatively short period but typical of Arctic-breeding geese.

Status. Even prior to 1900 it was recognized that barnacle geese occasionally visit the eastern United States. Bent (1925) summarized these early records, which were mostly for October and November, and extended from Vermont through Massachusetts, Long Island, and North Carolina. Godfrey (1966) likewise summarized early and more recent records for Canada, which included specimens from Baffin Island and Quebec and sight records for Labrador and Ontario. Recent North American records from eBird include Canadian records from Ontario, Quebec, New Brunswick, and Nova Scotia. US eBird records include Maine, New Hampshire, Vermont, New York, Massachusetts, Rhode Island, Connecticut,

Barnacle goose, adult with brood

Rhode Island, New Jersey, Pennsylvania, Delaware, Maryland, Virginia, North Carolina, South Carolina, and Florida There are also older records from Wisconsin, Illinois, Tennessee, Arkansas, Nebraska, Texas, and New Mexico.

Ogilvie and Boyd (1976) summarized population estimates of barnacle geese wintering in Britain and reported that the Greenland component numbered about 24,000 birds in 1973, an increase from 8,300 in 1959. The population breeding on Spitsbergen (Svalbard) also winters in Scotland; in 1962 it numbered about 2,300 birds, and was estimated at about 4,000 individuals by 1970. The population breeding on Novaya Zemlya and Vaigach Island consisted of about 20,000 individuals in 1959 and about 25,000 by 1970. All of these figures would suggest a total 1970 world population of about 50,000 barnacle geese (Kumari, 1971). More recently, Kear (2005) estimated breeding populations of

Barnacle goose, adult landing

32,000 in Greenland, 120,000 in Russia, 12,700 in Svalbard, and 12,000 in Sweden. Probably all the North American occurrences come from the Greenland population, which winters mainly in western Scotland (Inner and Outer Hebrides) and Ireland.

Relationships. Plumage and behavioral similarities suggest that the barnacle goose and Canada goose are close relatives (Johnsgard, 1965). Ploeger (1968) discussed the evolutionary history of the barnacle goose from the perspective of Pleistocene glaciation patterns.

Suggested readings. Ferns and Green, 1975; Jackson et al., 1974; Owen, 1978, 1980; Kear, 2005.

Brant

Branta bernicla (Linnaeus) 1758

Other vernacular names. American brant, black brant, brent (UK), brant goose, dark-bellied brant, light-bellied brant, Pacific brant

Range. Circumpolar, breeding along Arctic coastlines of North America and Eurasia as well as on Greenland, Iceland, and other high Arctic islands. In North America, winters on coastal areas south on the Pacific coast from British Columbia to northwestern Mexico and on the Atlantic coast from New England to North Carolina. Some wintering of North American breeders also occurs in Europe.

North American subspecies. *B. b. hrota* (Muller): Light-bellied (Atlantic) brant. In North America, breeds in Canada's Foxe Basin (Baffin Island, Melville Peninsula, and Southampton Island); the population winters along the US Atlantic coast ("Atlantic brant"). Also breeds on northern and western Greenland and adjacent Queen Elizabeth Islands ("Eastern High Arctic Brant"); a population uniquely wintering in Ireland. Brant breeding on the westernmost Parry Islands (Melville, Prince Patrick) winter on Padilla, Samish, and Fidalgo Bays in coastal Washington and Boundary Bay, British Columbia ("Western High Arctic Brant").

B. b. nigricans (Lawrence): Black (Pacific) brant. In North America, breeds from Alaska's Yukon-Kuskokwim Delta north locally along the Seward Peninsula and on the North Slope (Prudhoe Bay, Colville River Delta) to the Anderson River Delta (Northwest Territories); also breeds along coastline and islands (Banks, Victoria) of the Queen Maud Gulf; winters from California to Baja California and along coastal Sonora and Sinaloa. Considered by Delacour (1954) to be part of the Pacific brant, *B. b. orientalis*, which winters in Korea and China, but all brant subspecies are still unsettled, pending further DNA analyses.

Measurements. Atlantic (light-bellied) brant: *Wing:* males, 338–348 mm, ave. of 3, 344 mm; females, 323–333 mm, ave. of 7, 329 mm (Cramp and Simmons, 1977). *Culmen:* males 32–36mm, ave. of 12, 32 mm; females 30–34 mm, ave. of 8, 32 mm (Palmer, 1976).

Black (Pacific, dark-bellied) brant: *Wing:* males 325–344, ave. of 12, 335.5 mm; females 319–361, ave. of 8, 332 mm (Palmer, 1976). *Culmen:* ave. of 152 breeding males, 33.9 mm; ave. of 278 adult females, 32.2 mm (Reed, Ward, Derksen, and Sedinger, 1998).

Black brant, adults in flight

Weights (mass). Atlantic (light-bellied) brant: Ave. of 19 males, 3.4 lb. (1,542 g), max. 4.0 lb. (1,814 g). Ave. of 14 females, 2.8 lb. (1,270 g), max. 3.9 lb. (1,759 g.) (Nelson and Martin, 1953). Ave. of 17 males, 1,165 g; ave. of 152 females, 1,025 g (Reed, Ward, Derksen, and Sedinger, 1998).

Black (Pacific, dark-bellied) brant: Males, ave. of 26, 3.4 lb. (1,542 g), max. 4.9 lb. (2,223 g). Females, ave. of 15, 3.1 lb. (1,406 g), max. 3.6 lb. (1,633 g) (Nelson and Martin, 1953). Ave. of 189 males, 3.19 lb. (1,447 g), max. 4.0 lb. (1,814 g). Females, ave. of 181, 2.87 lb. (1,302 g), max. 3.81 lb. (1,724 g) (Hansen and Nelson, 1957). Ave. of 48 males, 1,165 g; ave. of 152 breeding females, 1,025 g (Reed, Ward, Derksen, and Sedinger, 1998).

Identification

In the hand. The brant's tiny mass (usually under four pounds, or 1,800 grams) will separate this species from all others except cackling geese, which have white on their cheeks instead of on the upper neck. Additionally, the central tail feathers of cackling geese extend beyond the tip of the tail coverts, which is not true of the brant goose.

In the field. In their coastal habitat, brant are usually seen in small flocks on salt water some distance from shore, their white hindquarters higher out of the water than is typical of ducks. The head, neck, and breast appear black, the sides grayish to whitish. Brant swim with their hindquarters well raised, exposing the white tail-coverts. When in flight, the birds appear short-necked, and the white hindquarters contrast strongly with the black foreparts, while both the upper and lower wing surfaces appear grayish brown. The birds usually fly in undulating or irregular lines rather than in V-formations like Canada or cackling geese. Compared to cackling geese, they have surprisingly soft and guttural notes, *r-r-r-ruk* or *ruk-ruk*.

Age and Sex Criteria

Sex determination. No plumage differences are available for external sex determination.

Age determination. *Adults* of both sexes have the head, neck, and upper breast black, except for a somewhat crescent-shaped white patch on both sides of the upper neck, interrupted by parallel striations of the neck feathers, exposing the darker bases. The back and scapulars are dark brown, with inconspicuous lighter tips, and the rump is dusky centrally and white laterally. The tail coverts are white and elongated, nearly hiding the black tail, and the abdomen is whitish posteriorly, and white or gray anteriorly. The sides and flanks vary from nearly white (*hrota*) to black and white (*nigricans*), with some blackish barring usually evident toward the rear. The flight feathers are black, and the upper wing coverts are grayish brown to dark brown, with inconspicuous paler edging. The bill, legs, and feet are black. *Immatures* in their first winter have conspicuous white edges on the upper wing-coverts, scapulars, and back. At least one or more white-tipped secondary coverts will also identify yearling birds until their summer flightless period, according to Harris and Shepherd (1965), who also found that at least some females

Map 8. Breeding (inked) and wintering (shaded) distributions of the brant in North America. Major breeding population groups are enclosed by dashed lines: BB = Black Brant, WHAB = Western High Arctic Brant, EHAB = Eastern High Arctic Brant, AB = Atlantic Brant. In part after Baldassarre (2014).

apparently breed at two years of age. Yearling males have penile development ranging from the typical small and unsheathed juvenile condition to the full adult condition, while all older age classes of males have a fully adult penis condition.

Distribution and Habitat

Breeding distribution and habitat. In Alaska, the black brant breeds abundantly in the Yukon-Kuskokwim Delta and in smaller numbers northward on the Seward Peninsula and along northern Kotzebue Sound, and eastward across Alaska's northern coastline. In Canada, the black brant extends from the Alaska border north to Banks Island and east along the Arctic coast in the Queen Maud Gulf to Bathurst Inlet and Perry River. Black brant (*orientalis*) also breed in the Bering Sea on St. Lawrence Island and in coastal northeastern Russia.

From Perry River and King William Island eastward, the black brant is replaced by the Atlantic brant. To the northeast, the Eastern High Arctic group of the Atlantic brant breeds from the Adelaide Peninsula northeast to Prince Charles Island and Bylot Island (off Baffin Island) and along the western coast of Baffin Island south to the Foxe Peninsula and (in Hudson Bay) Southampton Island. To the northwest, the Western High Arctic group of the Atlantic brant population breeds in the Parry Islands (Prince Patrick, Eglington and Melville Islands). In the far north, the Eastern High Arctic group of the Atlantic brant breeds on Elef Ringnes, Alex Heiberg, and Ellesmere Islands (Queen Elizabeth Islands) and northern Greenland, making them probably the most Arctic-nesting of the North American geese.

The typical breeding habitat of brant geese is lowland coastal tundra, usually just above high tide line, which makes the nesting grounds highly susceptible to flooding by storm tides. Low islands of tundra lakes and dry inland slopes well covered with vegetation are used to some extent as well. In the Yukon-Kuskokwim Delta, the heart of the black brant's breeding region, nesting occurs on low, grass-covered flats dissected by numerous tidal streams in a belt 2 to 3 miles wide (Spencer et al., 1951). In this area the brant prefer the short sedge cover, with the highest nest density (up to 144 per square mile reported) found 3 to 5 miles from the coast (Hansen and Nelson, 1957). However, at Prince Patrick Island, at the northern edge of the breeding range, brant nest on grassy mountain slopes up to 3 miles inland and usually at least a mile from the coast. Nest densities there are much lower, with a dozen pairs scattered over several square miles, and the nests are situated several hundred yards apart (Handley, 1950).

Wintering distribution and habitat. According to mid-twentieth-century midwinter survey averages, slightly more than half of the brant population wintered on the Atlantic coast, while the remainder occurred on the Pacific coast from southern British Columbia to the Pacific coast lagoons of Baja California, plus the coastal lagoons of Sonora and Sinaloa, Mexico. Since the 1950s there has been a major shift of Pacific coast birds wintering southward, so that the vast majority now (2016) use the coastal lagoons of Baja California. There are occasional inland records of brant to Texas, Oklahoma, Nebraska, and South Dakota.

On the west coast, the preferred wintering habitat of black brant consists of large areas of shallow marine water covered with eelgrass (*Zostera marina*), which is usually to be found in bay situations. In 1952 a total wintering population inventory revealed about 175,000 birds, 63 percent of which occurred in Baja California (Leopold and Smith, 1953). At that time California accounted for 25 percent of the total (mostly in Humboldt Bay, Morro Bay, and bays in Marin County), Washington supported 9 percent (mostly in Puget Sound), and the remaining 3 percent were distributed along the coasts of Oregon, British Columbia, and southeastern Alaska. Smith and Jensen (1970) also reported on Mexico's wintering brant population and documented a major shift from traditional wintering areas to coastal Sonora and Sinaloa, where more than 35,000 birds wintered in 1969. During the 2001–08 period, an average of 122,163 brant wintered in Mexico, mostly on western coastal lagoons of Baja California, with 85 percent of all black brant surveyed from 1998 to 2007 being widely distributed (from north to south) on San Quintín, Scammon, San Ignacio, and Magdalena lagoons.

On the Atlantic coast, most wintering by light-bellied brant occurs from Massachusetts south, with occasional records from the interior as far west as Illinois. Shallow expanses of salt water on coastal bays are prime habitat, with the birds in the Chesapeake Bay area being most abundant along the barrier beach side of the bays, concentrated wherever sea lettuce (*Ulva lactuca*) is abundant. Along the eastern Chesapeake, they concentrate in Tangier Sound and adjoining estuaries, especially where eelgrass and widgeon grass (*Ruppia*) are commonly found, and sometimes also occur in shallow areas of brackish water. The Chesapeake Bay flock represented about 5 percent of the Atlantic coast population around 1960 (Stewart, 1962), which is almost entirely restricted in winter to coastal areas from Massachusetts to North Carolina. Brant sometimes occur during migration on the lower Great Lakes (Sheppard, 1949) but generally are restricted as regular wintering birds to saltwater habitats.

General Biology

Age at maturity. Harris and Shepherd (1965) reported that 6 of 19 black brant that they examined had evidently nested as two-year-olds, but that no yearlings showed any signs indicating breeding. Barry (1967) estimated that possibly 10 percent of two-year-olds might breed during favorable nesting seasons.

Pair-bond pattern. Presumably the usual strongly monogamous pair-bond pattern typical of all geese applies to brant as well; at least no observations of wild or captive birds contradict this view. Einarsen (1965) noted that when Atlantic brant pairs occur on the Pacific coast the pairs are generally inseparable, and he suggested that the reason for the lack of interbreeding on Prince Patrick Island, where both light-bellied and black brant reportedly occur, is that strong pair-bonds are formed prior to arrival at the breeding grounds, and thus mixed pairing rarely, if ever, happens. This situation needs further study.

Nest location. As noted earlier, nests are usually located in low herbaceous vegetation often close to the high tide line but sometimes in upland situations. Barry (1956) mentioned that the majority of the nests he observed in a colony on Southampton Island were on small river delta islands covered with low, thick grass and were less than a mile from the coast. Later (1962) he stated that preferred nesting habitat is covered with sedge mat vegetation extending only about one-quarter mile inland from the normal high tide line and that more than 90 percent of the brant nested in this zone. Einarsen (1965) stated that small islands only a few square feet in area, or a small promontory extending out into a pond or lake, are often selected. The nests are usually bowl-shaped, and thus a sitting goose is able to flatten out on the nest so as to be barely visible above ground level.

Black brant, adults swimming

Clutch size. Barry (1956) reported that clutch sizes in a colony containing 203 black brant nests varied in different areas from 3.77 to 4.41, with an overall average of 4.0 and an observed range of 1 to 7 eggs. Hansen and Nelson (1957) reported that 116 black brant nests had an average clutch of 3.5 eggs, and Gillham (cited by Einarsen, 1965) noted that in 1939 a sample of 83 black brant nests averaged 4.96 eggs per clutch, while in 1940 a total of 108 nests averaged 3.8 eggs per clutch. This reduction in clutch size was evidently related to a severe freeze occurring about 8 days after migration had terminated.

Barry (1962) found a strong relationship between weather and clutch size. In the favorable 1953 season he found an average clutch of 4.6 eggs in 13 completed nests that had not yet suffered any egg losses. In two seasons that were retarded by cold weather, not only were the average clutch sizes smaller (4.3 for 109 completed clutches and 3.9 for 33 completed clutches) but also the nests that were started late had smaller clutches than the earliest ones. The collective average clutch size for 853 nests was 3.94 eggs.

The eggs are generally laid at the rate of 1 per day, but frequently a day might be skipped toward the end of the egg-laying period (Barry, 1956). At least one case of attempted renesting following freezing weather has been reported (Gillham, in Einarsen, 1965). Barry (1962) also mentioned that a few cases of attempted renesting occurred in the colony he studied, but the clutches were not successfully completed.

Incubation period. Barry (1956, 1967) reported the incubation period to be 24 days for 10 of 12 black brant nests, with one case each of 23 and 25 days, which is similar to a reported average of 23 days for 7 nests (Reed et al., 1998). Another estimate of the incubation length was 22 to 25 days (Mickelson, 1975).

Fledging period. Barry (1967) indicated that 40 to 45 days are required for black brant to attain flight. Mickelson (1975) reported a similar 40-day fledging period for black brant.

Nest and egg losses. Because of the vulnerability of brant nests to flooding, nest and egg losses are likely to be high in some years or in certain locations. Barry (1962) noted that predation and other losses took 27 percent of 723 black brant eggs in marked nests during three years of study. During three years in the Kashunuk River study area of the Yukon-Kuskokwim Delta, the hatching success of black brant nests ranged from 81 to 85 percent (US Fish and Wildlife Service, 1962).

Specific data on flooding effects are not available, but Einarsen (1965) mentioned that the Yukon-Kuskokwim Delta had severe floods and storms in 1952 and 1963, with resultant high nest and brood losses. The 1963 storm, associated with high tides, flooded nearly the entire brant nesting zone in the Clarence Rhode National Wildlife Range (now part of Yukon Delta National Wildlife Refuge), destroying thousands of eggs and young brant. However, Jones (1964) found that the percentage of immature brant seen during fall counts in Izembek Bay was sufficiently high (23 percent) to indicate that production in other areas was adequate to offset this localized loss.

Burton (1960) concluded from age-ratio counts of brant in Europe that the 1958 breeding season in the Soviet Arctic was associated with abnormally low temperatures during the summer months and must have been nearly a complete failure. More recently, light-bellied brant from the Eastern High Arctic Population had two disastrous production years in 2012 and 2013, with juveniles composing less than 2 percent of the fall population—in good years the percentage may be as high as 20 to 30 (Canadian Wildlife Service Waterfowl Committee, 2014.)

Black brant, adult threat posture

More recent counts by Jones (1970) indicated that the annual average incidence of juveniles in fall brant populations ranged from 18 to 40 percent between 1963 and 1969. Family groups contained averages of 2.58 to 2.86 juveniles. The percentage of apparent nonbreeders ranged from 31 to 69 percent over a 4-year period, averaging 56 percent. During these years the fraction of juveniles averaged 25 percent; thus a nonbreeding or unsuccessfully breeding segment of about 50 percent of adult-plumaged birds is apparently typical, even during years of good reproduction.

Juvenile mortality. Studies in Alaska (US Fish and Wildlife Service, 1962) indicated that the average brood size of first-week young ranged from 3.4 to 3.8 goslings in 3 different years. By 3 weeks of age the average brood size had been reduced to 2.2 to 3.2 birds. Finally, counts of juveniles in fall flocks at Izembek Bay suggest average family sizes of 2.58 to 2.86 juveniles per successful pair (Jones, 1970). Ignoring pairs that completely lost their eggs or young, it is evident that about half of the hatched young are often lost before reaching the wintering grounds. These first-year birds are more vulnerable than adults to various kinds of mortality; Hansen and Nelson (1957) estimated an average annual mortality rate of 45.4 percent for juveniles, based on direct recoveries of birds banded in Alaska.

Adult mortality. Kirby et al. (1986) calculated an average adult survival rate of 79 percent (range 59–98 percent) for Atlantic brant banded between 1956 and 1975). Hansen and Nelson (1957) likewise estimated an adult annual mortality rate of 21.8 percent (79.2 percent survival) on the basis of direct recoveries of birds banded as adults. If indirect recoveries through the sixth year are added, the estimated annual adult mortality rate was 32.2 percent. Boyd (1962) recalculated these figures and concluded that a mean adult mortality rate of 15 percent (85 percent survival) was typical of this population, compared to a 14 percent mortality rate (86 percent survival) of brant wintering in Britain, and 17 percent mortality (83 percent survival) for birds breeding on Spitsbergen. Because of these high survival rates, brant tend to be long-lived, with reported longevities up to nearly 26 years for males, nearly 22 years for females (Reed et al, 1998), and 28.2 years for a UK-banded bird (Kear, 2005). Three different estimates of annual survival rates for Alaska females ranged from 85 to 90 percent, a remarkable statistic considering the species' small size, the population's long migrations (often at least 2,000 miles to northwestern Mexico), and the difficulties of breeding under Arctic conditions.

Food and foraging. The close relationship between the distribution and abundance of eelgrass (*Zostera marina*) and brant has long been recognized (Cottam et al., 1944). Second in importance to eelgrass, and used by brant when eelgrass is absent or depleted by disease, is sea lettuce (*Ulva* species, especially *U. lactuca*). In some areas widgeon grass (*Ruppia*) is used to a limited extent by Atlantic brant (Martin et al., 1951). However, eelgrass is clearly the preferred food of both the Atlantic and Pacific populations, and the most extensive eelgrass beds in the world occur in Izembek Bay, Alaska, which during the fall temporarily supports the entire Pacific coast brant population (Jones, 1964; Jones and Jones, 1966). Here, about a quarter million brant historically fed on about 40,260 acres of eelgrass lying just below the water surface or exposed during low tide. The leaves of the eelgrass form dense mats often arranged in windrows, along which the brant swim while feeding. The protein content of the eelgrass in this bay averages about 7 percent, whereas samples of eelgrass and sea lettuce from Washington and Oregon average about 15 percent protein (Einarsen, 1965).

Sociality, densities, territoriality. As might be expected in a species that nests in a colonial fashion, brant geese are relatively social and gregarious. Jones and Jones (1966) reported seeing little strife in flocks consisting of two or three family groups, but hostile encounters were common in larger fall flocks. These generally were initially related to maintaining the integrity of family groups. However, by early October hostile encounters between adults and juveniles indicated that family bonds were being broken. This dissolution of family bonds was completed by late October, after which hostile encounters were again rarely seen, and the population consisted of a few very large groups containing all age groups. Thus, unlike most geese, family bonds are evidently not maintained through the first winter of life. Einarsen (1965) has also emphasized the gregariousness of brant geese, noting their strong tendency to "raft" in clusters and breed in colonies.

Estimates of breeding densities have been made by Hansen and Nelson (1957), who noted that in the best nesting areas on the Yukon-Kuskokwim Delta nest densities were as high as 144 per square mile and occur in the short sedge zone 3 to 5 miles from the coast. Barry (1956) reported a colony of about 700 nesting pairs in a stretch of coast about 4.5 miles long and usually less than 0.25 mile wide, or a little more than a square mile of area. Nesting density within this area varied considerably, with the highest density on the islands of the Boas River Delta. The distribution of these islands, about a foot above high tide, evidently strongly affected the breeding density. Thus, territoriality probably plays only a minor if not negligible role in affecting brant nesting densities. Territories are maintained by bluffing rather than fighting, according to Berry, but sometimes birds would be chased off a nesting territory and "escorted" away for some distance.

Interspecific relationships. Barry (1956) noted that the black brant colony he studied on the Anderson River Delta was entirely separated from the colony of snow geese, which nested on higher ground at least 0.25 mile inland from the high tide line. In Alaska the black brant nests in association with cackling geese on the Yukon-Kuskokwim Delta, but brant nesting occurs slightly closer to the coast (Spencer et al., 1951). Major avian predators of eggs are gulls and jaegers, especially the parasitic jaeger, but the Arctic fox often causes heavy destruction to nesting colonies and probably is primarily responsible for the brant's tendency to select coastal or delta islands for its nesting sites (Barry, 1967). Arctic foxes are also serious predators in the Yukon-Kuskokwim Delta, and their removal can significantly improve nesting success (Anthony, Flint, and Sedinger, 1991).

General activity patterns and movements. Einarsen (1965) commented on the brant's unusual flying ability, noting that he had clocked flying brant at a ground speed of 62 miles per hour, compared with 36 to 40 miles per hour for Canada geese. He also noted the brant's faster wing beats (3–4 per second) and more streamlined body and wings. When facing a headwind, the birds almost skim the wave tops, but even without strong headwinds the birds do not fly high. Flocks usually are arranged in long strings, undulating somewhat in flight, in marked contrast to the more highly organized flight formations of larger geese.

According to Einarsen, brant rarely reach a height of more than 200 to 300 feet above the ocean and certainly rarely stray far from salt water. Lewis (1937) has, however, described the probable migration route across western Quebec between Hudson Bay and the brant's wintering grounds on the Atlantic coast. The distance from the Gulf of St. Lawrence to James Bay or Ungava Bay is nearly 600 miles, apparently made in a nonstop flight at considerable height during the nighttime hours. Birds arriving at the mouth of the St. Lawrence River in late spring are typically in small flocks, with 155 such flocks averaging about 40 birds and rarely exceeding 100 birds (Lewis, 1937). However, when the brant arrive at the nesting grounds the flocks seldom exceed 20, and the birds are mostly paired (Barry, 1956).

Social and Sexual Behavior

Flocking behavior. The generally gregarious nature of brant geese has already been mentioned. Even shortly after hatching their broods, families will sometimes merge. Barry (1956) mentioned seeing 6 to 8 adult black brant with 10 to 14 young swimming in a group. Einarsen (1965) illustrated 2 pairs of adults with 9 to 10 very recently hatched young being closely convoyed. Typically a family consists of an adult swimming ahead, the young birds next, and the other adult taking up the rear, a trait that persists almost until the families break up in the fall (Jones and Jones, 1966).

Black brant, adult leading brood

Flocking by nonbreeding birds is also typical of brant; Barry (1956) noted that yearling birds remained separate from the nesting colony on Southampton Island. This group of about 200 birds flew out to feed each day in flocks of 40 to 50 birds, and during the midsummer molt they congregated in a bay by the Boas River.

Pair-forming behavior. Little has been written on pair-forming behavior in brant, but permanent pair-bonding is present (Owen, 1980). Barry (1956) observed two instances of possible courtship flights involving three birds but noted that nearly all the birds were mated prior to arrival at their breeding grounds. Einarsen (1965) also mentioned seeing several trios of seemingly courting brant on their wintering grounds between mid-January and late March. He believed that the female took the lead in these flights and was followed by two or more competing males. Among ducks, similar three-bird flights typically consist of a pair and an intruding male and can often be seen in wintering snow goose flocks. It is highly probable that pair-bonds are formed and maintained in brant, as in other geese, by the repeated performance of the triumph ceremony between two birds. Such ceremonies can be seen between paired birds in captive flocks. Under wild conditions, pairing most probably occurs during the second year of life, and a year or two before sexual maturity (Barry, 1962). Jones and Jones (1966) mentioned seeing apparent hostility postures between wild birds that met and moved away together, and noted that this behavior frequently was the means by which a family member regained its own group.

Copulatory behavior. Precopulatory behavior in brant consists of mutual head-dipping movements that resemble bathing movements, but it lacks the strong tail-cocking elements present in many geese. At times this head-dipping also takes the form of upending and is followed by mounting. The precopulatory display consists of the male lifting his bill, stretching his neck, and calling, but neither sex exhibits wing-lifting or wing-spreading at this time (Johnsgard, 1965).

Nesting and brooding behavior. Observations on nesting behavior in black brant have been provided by Barry (1956). Pairs establish nesting territories as soon as the habitat is free of ice and snow. The nest is usually a simple hollow about 2 inches deep and 9 inches across, with a sparse amount of grass pulled up from the immediate vicinity of the nest. Down is deposited with the laying of the first egg and increased with each additional one. Females cover the eggs with down whenever they leave the nest, and during the egg-laying period the male remains within 100 yards of the nest. However, once incubation is underway, the female rests low and inconspicuously on the nest, with her neck extended along the ground. When she returns to the nest after feeding, the male escorts her until he is about 15 feet from the nest, and after the female is back on the nest he returns to a distance of about 50 to 100 yards away to forage and keep watch. At times the male might fly at gulls and jaegers, chasing them from the vicinity of the nest.

When the young are hatched, they do not remain on the nest long but are soon led out to the edges of the tidal flats, where they apparently feed on larvae and small crustaceans (Barry, 1956). Einarsen (1965) reported that the young also feed on the tender parts of sedges and, when disturbed, can effectively dive within a few days of hatching. Both sexes closely attend the young, feeding with them up to 14 hours per day in Alaska, although the adults become flightless about 7 to 10 days after the young are hatched. Gulls, jaegers, Arctic foxes, and polar bears are known predators, and in Alaska up to 20 percent of pairs may lose their entire broods (Kear, 2005). The flightless period lasts about 30 days; thus the adults are normally again able to fly about the time the young are fledged, but sometimes only shortly before the onset of freezing weather (Barry, 1962). The young fledge at 40 to 45 days, one of the shortest fledging periods of all geese, and in the high Arctic families may leave breeding areas as early as two weeks after fledging (Kear, 2005).

Post-breeding behavior. With the arrival of freezing weather shortly after fledging, the adults and young gradually move to more southerly areas. In the case of the Pacific coast population, this is the Izembek Bay area on the northwestern tip of the Alaska Peninsula. Fall arrival there averages about August 25, and a mass departure occurs about 8 weeks later (Jones, 1964). Lewis (1937) described the fall migration route of the birds wintering on the Atlantic coast and noted that the east side of Hudson Bay and

Ungava Bay are probably important fall staging areas.

King's (1970) observations of large numbers of molting black brant near Cape Halkett, Alaska, numbering perhaps as many as 25,000 birds, is of special interest. These congregations would suggest that birds breeding farther east or south in Canada might congregate there in nonproductive years, or that the Arctic Slope might support a greater breeding population of brant than had been previously believed.

Status. Bellrose (1976) reported that although the average winter population of Atlantic brant between 1955 and 1974 was 177,000 birds, a sharp decline reduced the number from 151,000 in 1971 to 42,000 in 1973, which was followed by an increase to nearly 90,000 during the two following years. This serious reduction was largely the result of some unfavorable nesting seasons but in part resulted from over-harvesting (Penkala et al., 1975). In comparison, the US Fish and Wildlife Service's 2015 midwinter survey index for Atlantic brant was 111,400 birds, 16 percent below the 2014 index and a continuation of a long-term average decline of 3 percent annually for the 2006–15 ten-year average. The percent of juveniles in the 2014 fall population was only 9.5 percent, or well below the ten-year average of 18 percent. Recreational hunting is typically closed in the United States when the population falls below 100,000 birds, as was true in the 1970s.

During the 2013 and 2014 regular hunting seasons, the total estimated brant hunting kills in the United States (excluding Alaska) averaged 13,709 (Raftovich, Chandler, and Wilkins, 2015). This average is substantially below the average of 24,088 shot during 2008 and 2009 seasons (Raftovich, Chandler, and Wilkins, 2010), and also fewer than the more than 35,000 killed in 2002. Most of the brant shot in the United States are Atlantic brant, with large numbers taken in New York, New Jersey, Virginia, and North Carolina (Baldassarre, 2014). Only a few hundred Atlantic brant are killed in Canada, counting both sport hunters and subsistence hunters.

The light-bellied brant ("Eastern High Arctic Brant" group) that mostly breed in the eastern and central Queen Elizabeth Islands undertake a transhemispheric migration to winter mostly along the coast of Ireland, where they join brant from other breeding regions. Their collective numbers were about 6,100 birds in 2001 (Denny et al., 2004), but total brant numbers had climbed to more than 41,000 in the Irish wintering grounds by 2012 (Canadian Wildlife Service Waterfowl Committee, 2014).

The black brant wintering along the Pacific coast of North America (the "Pacific Coast Brant," including the "Black Brant" and "Western High Arctic Brant" population groups) evidently remained at a fairly constant population level from 1951 to 1974, with an average wintering population estimate of 140,000 birds during that period (Bellrose, 1976). Based on midwinter ground surveys in Mexico plus aerial surveys in the United States, the US Fish and Wildlife Service's 2015 midwinter survey index for black brant (including the Western High Arctic Brant population) was 136,500 birds, or 21 percent lower than the 2014 estimate of 173,300. However, the ten-year 2006–15 population index indicated no significant trend had occurred over that period (US Fish and Wildlife Service, 2015).

Most black brant now winter too far south to be affected much by US sport hunters. The total annual recreation kill of black brant during the late 1980s was about 8,000 birds, with nearly half taken in Mexico, about 20 percent in the United States, 3 percent in Canada, and the remainder being sustenance kill by native hunters (Reed et al., 1998). A small number of black brant are shot annually during a short and late hunting season in British Columbia (Canadian Wildlife Service Waterfowl Committee, 2014). Several thousand black brant are killed annually by subsistence hunters in Alaska (Wentworth and Wong, 2001; Wolfe and Paige, 1995), as well as some in Canada and Russia, although estimated numbers are probably impossible to obtain.

A 2015 Canadian estimate of all four North American brant population groups in 2013 included 200,000 birds in the "Atlantic Brant Population" group. It was in a slightly declining ten-year trend, down from about 250,000 in the early 2000s and from an all-time high of nearly 290,000 in the 1960s. The Eastern High Arctic Population group was estimated at about 35,000 birds and was judged to be slowly increasing, from a low point of less than 10,000 in the 1960s. The Western High Arctic Population group was judged to be stable at about 16,000 birds. The Black Brant Population group was estimated to be about 150,000 birds in 2013 and judged increasing, although it had briefly reached nearly 200,000 in the early 1980s (Canadian Wildlife Service Waterfowl Committee, 2014; 2015).

Relationships. Ploeger (1968) has discussed the probable basis of subspeciation in brant in connection with glacial events; most writers now accept the idea that only a single species of brant should be recognized. However, this species is well distinguished from all of the other extant *Branta* species. Its downy young most closely resemble those of the barnacle goose, which is likewise a rather marine-adapted and high-Arctic species, and is perhaps its nearest living relative (Johnsgard, 1965). Livezey (1996) concluded from a variety of morphological data sources that the red-breasted goose (*Branta ruficollis*) of central Eurasia is the nearest living relative of brant.

Suggested reading. Barry, 1966; Mickelson, 1973; Ogilvie 1978; Owen, 1978, 1980; Madsen, Bregnballe, and Mehlum, 1989; Sedinger et al., 1993; Reed et al., 1998; Kear, 2005; Baldassarre, 2014.

Black brant, adult threat

II.

References

The Birds of North America Monographs of the American Ornithologists' Union and the Philadelphia Academy of Natural Sciences

Elly, C. R., and A. X. Dzubin. 1994. Greater White-fronted Goose. In *The Birds of North America*, No. 131. (A. Poole and F. Gill, eds.). Philadelphia, PA: The Academy of Natural Sciences, and Washington, DC: American Ornithologists' Union: The Birds of North America, Inc. 32 pp. (Includes about 350 citations.)

Mowbray, T. B., C. R. Ely, J. S. Sedinger, and R. E. Trost. 2003. Canada Goose. In *The Birds of North America*, No. 682. (A. Poole and F. Gill, eds.). Philadelphia, PA: The Academy of Natural Sciences, and Washington, DC: American Ornithologists' Union. The Birds of North America, Inc. (Includes about 400 citations.)

Mowbray, T. B., F. Cooke, and B. Ganter. 2000. Snow Goose. In *The Birds of North America*, No. 514. (A. Poole and F. Gill, eds.). Philadelphia, PA: The Academy of Natural Sciences, and Washington, DC: American Ornithologists' Union. The Birds of North America, Inc. 40 pp. (Includes about 225 citations.)

Petersen, M. R., J. A. Schmutz, and R. F. Rockwell. 1994. Emperor Goose. In *The Birds of North America*, No. 97. (A. Poole and F. Gill, eds.). Philadelphia, PA: The Academy of Natural Sciences, and Washington, DC: American Ornithologists' Union. The Birds of North America, Inc. 20 pp. (Includes about 80 citations.)

Reed, A., D. H. Ward, D. V. Derksen, and J. S. Sedinger. 1998. Brant. In *The Birds of North America*, No. 337. (A. Poole and F. Gill, eds.). Philadelphia, PA: The Academy of Natural Sciences, and Washington, DC: American Ornithologists' Union. The Birds of North America, Inc. 32 pp. (Includes about 200 citations.)

Ryder, J. P., and R. T. Alisauskas. 1995. Ross's Goose. In *The Birds of North America*, No. 162. (A. Poole and F. Gill, eds.). Philadelphia, PA: The Academy of Natural Sciences, and Washington, DC: American Ornithologists' Union. The Birds of North America, Inc. 28 pp. (Includes about 140 citations.)

General References

Alexander, S. A., R. S. Ferguson, and K. J. McCormick. 1991. *Key Migratory Bird Terrestrial Habitat Sites in the Northwest Territories.* Occasional Paper 71. Ottawa, ON: Canadian Wildlife Service.

American Ornithologists' Union (AOU). 1998. *The AOU Check-list of North American Birds.* 7th ed. Washington, DC: American Ornithologists' Union.

Anderson, M. G., and L. G. Sorenson. 2001. Global climate change and waterfowl: adaptation in the face of uncertainty. *Transactions of the North American Wildlife and Natural Resources Conference* 66: 300–319.

Anderson, M. G., J. M. Rhymer, and F. C. Rohwer. 1992. Philopatry, dispersal, and the genetic structure of waterfowl populations. Pp. 365–395 in B. D. J. Batt et al., eds. *Ecology and Management of Breeding Waterfowl.* Minneapolis: University of Minnesota Press.

Arctic Goose Joint Venture Technical Committee. 2008. *Arctic Goose Joint Venture Strategic Plan: 2008–2012.* Arctic Goose Joint Venture Coordination Office, Canadian Wildlife Service, Edmonton, AB.

Armstrong, T. 1996. Effects of research activities on nest predation in Arctic-nesting geese. *Journal of Wildlife Management* 60: 265–269.

Baker, A. J. 2007. Molecular advances in the study of geographic variation and speciation in birds. Pp. 18–29 in: Cicero, C., and J. V. Remsen, Jr. 2007. *Festschrift for Ned K. Johnson: Geographic Variation and Evolution in Birds.* Ornithological Monographs, No. 63. Washington, DC: American Ornithologists' Union.

Baldassarre, G. A, 2014. *Ducks, Geese, and Swans of North America.* Baltimore, MD: Johns Hopkins University Press.

Baldassarre, G. A., and E. G. Bolen. 2006. *Waterfowl Ecology and Management.* 2nd ed. Malabar, FL: Krieger Publishing.

Barry, T. W. 1967. Geese of the Anderson River Delta, Northwest Territories. PhD dissertation, University of Alberta, Edmonton, AB.

Batt, B. D. J., A. D. Afton, M. G. Anderson, C. D. Ankney, D. H. Johnson, J. A. Kadlec, and G. L. Krapu, eds. 1992. *Ecology and Management of Breeding Waterfowl*. Minneapolis: University of Minnesota Press.

Bellrose, F. C. 1980. *Ducks, Geese, and Swans of North America*. 3rd ed. Washington, DC: Wildlife Management Institute.

Bent, A. C. 1923. *Life Histories of North American Wild Fowl. Part 1*. US National Museum Bulletin 126. Washington, DC: US Government Printing Office.

Bent, A. C. 1925. *Life Histories of North American Wild Fowl. Part 2*. US National Museum Bulletin 130. Washington, DC: US Government Printing Office.

Brandt, H. 1943. *Alaska Bird Trails*. Cleveland, OH: Bird Research Foundation.

Burleigh, T. D. 1944. *The Bird Life of the Gulf Coast Region of Mississippi*. Louisiana State University, Museum of Zoology Occasional Papers, No. 20: 329–490.

Byrd, G. V., D. L. Johnson, and D. D. Gibson. 1974. The birds of Adak Island, Alaska. *Condor* 76: 288–300.

Campbell, R. W., N. K. Dawe, I. McTaggart-Cowan, J. M. Cooper, G. W. Kaiser, and M. C. E. McNall. 1990. *The Birds of British Columbia, Vol. 1: Nonpasserines, Introduction and Loons through Waterfowl*. Vancouver: University of British Columbia Press.

Canadian Wildlife Service Waterfowl Committee. 2013. *Population Status of Migratory Game Birds in Canada*. Canadian Wildlife Service Migratory Birds Regulatory Report No. 40. Ottawa, ON: Canadian Wildlife Service. http://www.ec.gc.ca/rcom-mbhr/default.asp?lang=En&n=B2A654BC-1

Canadian Wildlife Service Waterfowl Committee. 2014. *Population Status of Migratory Game Birds in Canada— November 2014*. Canadian Wildlife Service Migratory Birds Regulatory Report. Ottawa, ON: Canadian Wildlife Service. 230 pp. https://ec.gc.ca/rcom-mbhr/default.asp?lang=En&n=2C549580-1

Canadian Wildlife Service Waterfowl Committee. 2015. *Population Status of Migratory Game Birds in Canada— November 2015*. Canadian Wildlife Service Migratory Birds Regulatory Report. Ottawa, ON: Canadian Wildlife Service. https://www.ec.gc.ca/rcom-mbhr/default.asp?lang=En&n=9DB378FC-1

Conant, B., and R. J. King. 2006. *Winter Waterfowl Survey: Mexico West Coast and Baja California*. Juneau, AK: US Fish and Wildlife Service.

Conover, H. B. 1926. Game birds of the Hooper Bay region of Alaska. *Auk* 43: 162–180.

Cramp, S., and E. L. Simmons, eds. 1977. *Handbook of the Birds of Europe, the Middle East, and North Africa: The Birds of the Western Palearctic, Vol. 1. Ostriches to Ducks*. Oxford, UK: Oxford University Press.

Delacour, J. 1954. *The Waterfowl of the World*. Vol. 1. London: Country Life Ltd.

Del Hoyo, J., A. Elliott, and J. Sargatal, eds. 1994. *Handbook of the Birds of the World*. Vol. 1. Barcelona: Lynx Edicions.

Denson, E. P., Jr., and W. W. Bentley. 1962. The migration and status of waterfowl at Humboldt Bay, California. *Murrelet* 43: 19–28.

Derksen, D. V., M. W. Weller, and W. D. Eldridge. 1979. Distributional ecology of geese molting near Teshekpuk Lake, National Petroleum Reserve, Alaska. Pp. 189–207 in: R. L. Jarvis and J. C. Bartonek, eds. *Management and Biology of Pacific Flyway Geese*. Corvallis, OR: OSU Book Stores.

Derksen, D. V., T. C. Rothe, and W. D. Eldridge. 1981. *Use of Wetland Habitats by Birds in the National Petroleum Reserve, Alaska*. Resource Publication 141. Washington, DC: US Fish and Wildlife Service.

Ely, C. R., and K. T. Scribner. 1994. Genetic diversity in Arctic-nesting geese: implications for management and conservation. *Transactions of the North American Wildlife and Natural Resources Conference* 59: 91–110.

Erskine, A. J. 1992. *Atlas of Breeding Birds of the Maritime Provinces*. Halifax, NS: Nimbus Publishing and Nova Scotia Museum.

Gilligan, J., M. Smith, D. Rogers, and A. Contreras. 1994. *Birds of Oregon: Status and Distribution*. McMinnville, OR: Cinclus Publications.

Godfrey, W. E. 1986. *The Birds of Canada*. Rev. ed. National Museums of Canada, Bulletin No. 203, Biological Series No. 73. 428 pp.

Grinnell, J., and A. H. Miller. 1944. *The Distribution of the Birds of California*. Cooper Ornithological Club, Pacific Coast Avifauna, No. 27.

Grinnell, J., H. C. Bryant, and T. I. Storer. 1918. *The Game Birds of California*. Berkeley: University of California Press.

Harper, F. 1958. *Birds of the Ungava Peninsula*. Lawrence: University of Kansas Museum of Natural History Miscellaneous Publication No. 17

Hodges, J. I., D. J. Groves, and B. P. Conant. 2008. Distribution and abundance of waterbirds near shore in southeast Alaska, 1997–2002. *Northwestern Naturalist* 89: 85–96.

Hodges, J. I., J. G. King, B. Conant, and H. A. Hanson. 1996. *Aerial Surveys of Waterbirds in Alaska, 1957–94: Population Trends and Observer Variability*. Information and Technology Report 4. Juneau, AK: National Biological Service, US Department of the Interior.

Hohman, W. L., C. D. Ankney, and D. H. Gordon. 1992. Ecology and management of postbreeding waterfowl. Pp. 128–189 in: B. D. J. Batt et al., eds. *Ecology and Management of Breeding Waterfowl*. Minneapolis: University of Minnesota Press.

Jehl, J. R., Jr., and B. S. Smith. 1970. *Birds of the Churchill Region, Manitoba*. Special Publication No.1. Winnipeg: Manitoba Museum of Man and Nature. 87 pp.

Johnsgard, P. A. 1961. The taxonomy of the Anatidae—a behavioural analysis. *Ibis* 103a: 71–85. http://digitalcommons.unl.edu/johnsgard/29

Johnsgard, P. A. 1965. *Handbook of Waterfowl Behavior*. Ithaca, NY: Cornell University Press. 378 pp. http://digitalcommons.unl.edu/bioscihandwaterfowl/7/

Johnsgard, P. A. 1968. *Waterfowl: Their Biology and Natural History*. Lincoln: University of Nebraska Press. 138 pp.

Johnsgard, P. A. 1975. *Waterfowl of North America*. Bloomington: Indiana University Press. 573 pp. http://digitalcommons.unl.edu/biosciwaterfowlna/1 (rev. ed.)

Johnsgard, P. A. 1978. *Ducks, Geese, and Swans of the World*. Lincoln: University of Nebraska Press. 400 pp. http://digitalcommons.unl.edu/biosciducksgeeseswans/

Johnsgard, P. A. 1979. *A Guide to North American Waterfowl*. Bloomington: Indiana University Press. 270 pp.

Johnsgard, P. A. 1987. *Waterfowl of North America: The Complete Ducks, Geese, and Swans* (species accounts text). Augusta, GA: Morris Publishing. 135 pp.

Johnsgard, P. A. 2012a. *Wetland Birds of the Central Plains: South Dakota, Nebraska, and Kansas*. Lincoln, NE: University of Nebraska–Lincoln DigitalCommons and Zea Books. http://digitalcommons.unl.edu/zeabook/8/

Johnsgard, P. A. 2012b. *The Nebraska Wetlands and Their Ecology*. Lincoln, NE: Division of Conservation and Survey, University of Nebraska School of Natural Resources. 202 pp.

Johnsgard, P. A. 2012c. *Wings over the Great Plains: Bird Migrations in the Central Flyway*. Lincoln, NE: University of Nebraska–Lincoln DigitalCommons and Zea Books, 249 pp. http://digitalcommons.unl.edu/zeabook/13/

Johnsgard, P. A. 2013. *The Birds of Nebraska*. Rev. ed. Lincoln, NE: University of Nebraska–Lincoln DigitalCommons and Zea Books. 140 pp. http://digitalcommons.unl.edu/zeabook/17/

Johnsgard, P. A. 2015. *Global Warming and Population Responses among Great Plains Birds*. Lincoln, NE: University of Nebraska–Lincoln DigitalCommons and Zea Books. 383 pp. http://digitalcommons.unl.edu/zeabook/26/

Kear, J. 2005. *Ducks, Geese, and Swans*. 2 vols. Oxford, UK: Oxford University Press. 910 pp.

Kenyon, K. W. 1961. Birds of Amchitka Island, Alaska. *Auk* 78: 305–326.

Kessel, B. 1988. *Birds of the Seward Peninsula, Alaska: Their Biogeography, Seasonality, and Natural History*. Anchorage: University of Alaska Press.

King, J. G., and D. V. Derksen. 1986. Alaska goose populations: Past, present, and future. *Transactions of the North American Wildlife and Natural Resources Conference* 51: 464–479.

King, J. G., and J. I. Hodges. 1979. A preliminary analysis of goose banding on Alaska's Arctic slope. Pp. 176–188 in: R. L. Jarvis and J. C. Bartonek, eds. *Management and Biology of Pacific Flyway Geese*. Corvallis, OR: OSU Book Stores.

Lepage, D., D. N. Nettleship, and A. Reed. 1998. Birds of Bylot Island and adjacent Baffin Island, Northwest Territories, Canada, 1979 to 1997. *Arctic* 51: 125–141.

Livezey, B. C. 1986. A phylogenetic analysis of Recent anseriform genera using morphological characters. *Auk* 103: 7734–7454.

Livezey, B. C. 1996. A phylogenetic analysis of the geese and swans (Anseriformes, Anserinae), including selected fossil species. *Systematic Zoology* 45: 415–450.

Livezey, B. C. 1997. A phylogenetic classification of waterfowl (Aves: Anseriformes), including selected fossil species. *Annals of the Carnegie Museum* 66: 457–496.

Lutmerding, J. A., and A. S. Love. 2011. *Longevity Records of North American Birds*. Version 2011.2. Laurel, MD: Bird Banding Laboratory, Patuxent Wildlife Research Center.

MacInnes, C. D., E. H. Dunn, D. H. Rusch, F. Cooke, and F. G. Cooch. 1990. Advancement of goose nesting dates in the Hudson Bay region, 1951–1986. *Canadian Field-Naturalist* 104: 295–297.

Madge, S., and H. Burn. 1988. *Waterfowl: An Identification Guide to the Ducks, Geese, and Swans of the World*. Boston: Houghton Mifflin. 298 pp.

Madsen, J., and C. E. Mortensen. 1987. Habitat exploitation and interspecific competition in moulting geese in Greenland. *Ibis* 129: 25–44.

Murie, O. J. 1959. *Fauna of the Aleutian Islands and Alaska Peninsula*. US Dept. of Interior, US Fish and Wildlife Service, North American Fauna, No. 61.

Oates, D. W., and J. Principato.1994. Genetic variation and differentiation of North American waterfowl (Anatidae). *Transactions of the Nebraska Academy of Sciences* 21: 127–145.

Ogilvie, M. A. 1978. *Wild Geese*. Berkhamsted, UK: T. and A. D. Poyser.

Owen, M. 1977. *Wildfowl of Europe*. London: Macmillan. 254 pp.

Owen, M. 1980. *Wild Geese of the World: Their Life History and Ecology*. London: B. T. Batsford.

Palmer, R. S., ed. 1976. *Handbook of North American Birds*. Vol. 2: Waterfowl, Part 1. New Haven, CT: Yale University Press. 519 pp.

Parmelee, D. F., and S. D. MacDonald. 1960. *The Birds of West-Central Ellesmere Island and Adjacent Areas*. National Museum of Canada Bulletin 169.

Parmelee, D. F., H. A. Stephens, and R. H. Schmidt. 1967. *The Birds of Southeastern Victoria Island and Adjacent Small Islands*. National Museum of Canada Bulletin 222.

Platte, R. M., and R. A. Stehn. 2009. *Abundance, Distribution, and Trend of Waterbirds on Alaska's Yukon-Kuskokwim Delta Coast, Based on 1988 to 2009 Aerial Surveys*. Division of Migratory Bird Management, US Fish and Wildlife Service, Anchorage, AK.

Portenko, L. A. 1981. *Birds of the Chukchi Peninsula and Wrangel Island*. Vol. 1. [translated from Russian]. New Delhi, India: Amerind Publishing.

Portenko, L. A. 1989. *Birds of the Chukchi Peninsula and Wrangel Island*. Vol. 2. [translated from Russian]. Washington, DC: Smithsonian Institution Libraries and National Science Foundation.

Prevett, J. P., H. G. Lumsden, and F. C. Johnson. 1983. Waterfowl kill by Cree hunters of the Hudson Bay Lowland, Ontario. *Arctic* 36: 185–192.

Raftovich, R. V., K. A. Wilkins, K. D. Richkus, S. S. Williams, and H. L. Spriggs. 2009. *Migratory Bird Hunting Activity and Harvest during the 2007 and 2008 Hunting Seasons*. Laurel, MD: US Fish and Wildlife Service.

Raftovich, R. V., K. A. Wilkins, K. D. Richkus, S. S. Williams, and H. L. Spriggs. 2010. *Migratory Bird Hunting Activity and Harvest during the 2008 and 2009 Hunting Seasons*. Laurel, MD: US Fish and Wildlife Service.

Raftovich, R.V., S. C. Chandler, and K.A. Wilkins. 2015. *Migratory Bird Hunting Activity and Harvest during the 2013– 14 and 2014–15 Hunting Seasons*. Laurel, MD: US Fish and Wildlife Service.

Raveling, D. G. 1978. The timing of egg laying by northern geese. *Auk* 95: 294–303.

Reeber, S. 2016. *Waterfowl of North America, Europe, and Asia*. Princeton, NJ: Princeton University Press. 656 pp.

Rich, T. C., et al., eds. 2004. *North American Landbird Conservation Plan*. Ithaca, NY: Partners in Flight and Cornell University Laboratory of Ornithology.

Robbins, M. B., and D. A. Easterla. 1992. *Birds of Missouri: Their Distribution and Abundance*. Columbia: University of Missouri Press.

Rockwell, R., K. Abraham, and R. Jefferies. 1996. Tundra under siege. *Natural History* 105: 20–21.

Rohwer, F. C. 1986. The adaptive significance of clutch size in waterfowl. PhD dissertation, University of Pennsylvania, Philadelphia, PA.

Rohwer, F. C. 1988. Inter- and intraspecific relationships between egg size and clutch size in waterfowl. *Auk* 105: 161–176.

Rohwer, F. C., and D. I. Eisenhauer. 1989. Egg mass and clutch size relationships in geese, eiders, and swans. *Ornis Scandinavica* 20: 43–48.

Rose, P. M., and D. A. Scott. 1997. *Waterfowl Population Estimates*. 2nd ed. Wetlands International Publication 44. Wageningen, The Netherlands: Wetlands International.

Ross, R. K., K. F. Abraham, T. R. Gadawski, R. S. Rempel, T. S. Gabor, and R. Maher. 2002. Abundance and distribution of breeding waterfowl in the Great Clay Belt of northern Ontario. *Canadian Field-Naturalist* 116: 42–50.

Rusch, D. H., and F. D. Caswell. 1997. Evaluation of the Arctic Goose Management Initiative. Pp. 120–122 in: B. D. J. Batt, ed. *Arctic Ecosystems in Peril: Report of the Arctic Goose Habitat Working Group*. Arctic Goose Joint Venture Special Publication. US Fish and Wildlife Service, Washington, DC, and Canadian Wildlife Service, Ottawa, ON.

Samelius, G., and R. T. Alisauskas. 2009. Habitat alteration by geese at a large Arctic goose colony: Consequences for lemmings and voles. *Canadian Journal of Zoology* 87: 95–101.

Sargeant, A. B., and D. G. Raveling. 1992. Mortality during the breeding season. Pp. 396–422 in: B. D. J. Batt, et al., eds. *Ecology and Management of Breeding Waterfowl*. Minneapolis: University of Minnesota Press.

Sauer, J. R., J. E. Hines, J. E. Fallon, K. L. Pardieck, D. J. Ziolkowski, Jr., and W. A. Link. 2014. *The North American Breeding Bird Survey, Results and Analysis, 1966–2013*. Version 01.30.2015. Laurel, MD: USGS Patuxent Wildlife Research Center (www.pwrc.usgs.gov). http://www.mbr-pwrc.usgs.gov/bbs/bbs.html

Sedinger, J. S. 1984. Protein and amino acid composition of tundra vegetation in relation to nutritional requirements of geese. *Journal of Wildlife Management* 48: 1128–1136.

Semenchuk, G. P., ed. 2007. *The Atlas of Breeding Birds of Alberta: A Second Look*. Edmonton: Federation of Alberta Naturalists.

Shepherd, P. E. K. 1955. *Migratory Waterfowl Studies Nesting and Banding, Selawik Area*. Federal Aid in Wildlife Restoration Project W-3-R-11. Juneau: Alaska Game Commission.

Sinclair, P. H., W. A. Nixon, C. D. Eckert, and N. L. Hughes, eds. 2003. *Birds of the Yukon Territory*. Vancouver: University of British Columbia Press.

Smith, A. G. 1971. *Ecological Factors Affecting Waterfowl Production in the Alberta Parklands*. Resource Publication 98. Washington, DC: US Bureau of Sport Fisheries and Wildlife.

Smith, A. R. 1996. *Atlas of Saskatchewan Birds*. Regina: Saskatchewan Natural History Society.

Snyder, L. L. 1957. *Arctic Birds of Canada*. Toronto, ON: University of Toronto Press.

Stewart, P. A. 1962. *Waterfowl Populations in the Upper Chesapeake Region*. Special Scientific Report—Wildlife 65. Washington, DC: US Fish and Wildlife Service.

Thompson, S. C., and D. G. Raveling. 1988. Nest insulation and incubation constancy of Arctic geese. *Wildfowl* 39:124–132.

Todd, F. S. 1996. *Natural History of the Waterfowl*. Vista, CA: Ibis Publishing.

Todd, W. E. C. 1963. *Birds of the Labrador Peninsula and Adjacent Areas*. Toronto, ON: University of Toronto Press.

US Fish and Wildlife Service. 1951. *Waterfowl Populations and Breeding Conditions—Summer 1950*. Special Scientific Report: Wildlife, No. 8. Washington, DC.

US Fish and Wildlife Service. 1955. *Waterfowl Populations and Breeding Conditions—Summer 1954*. Special Scientific Report: Wildlife, No. 27. Washington, DC.

US Fish and Wildlife Service. 1956. *Waterfowl Populations and Breeding Conditions—Summer 1955*. Special Scientific Report: Wildlife, No. 30. Washington, DC.

US Fish and Wildlife Service. 1962. *Waterfowl Status Report, 1962.* Special Scientific Report: Wildlife, No. 68. Washington, DC.

US Fish and Wildlife Service. 2015. *Waterfowl Population Status, 2014.* Laurel, MD: US Fish and Wildlife Service.

US Fish and Wildlife Service. 2015. *Waterfowl Population Status, 2015.* Laurel, MD: US Fish and Wildlife Service.

Uspenski, S. M. 1965. The geese of Wrangel Island. *Wildfowl Trust Annual Report* 16: 126–129.

Vacca, M. M., and C. M. Handel. 1988. Factors influencing predation associated with visits to artificial goose nests. *Journal of Field Ornithology* 59: 215–223.

van Horn, K. 1991. Habitat use and activity patterns of interior Alaskan waterbirds. MS thesis, University of Missouri–Columbia, Columbia, MO.

Wahl, T. R., B. Tweit, and S. G. Mlodinow. 2005. *Birds of Washington: Status and Distribution.* Corvallis: Oregon State University Press.

Wentworth, C., and D. Wong. 2001. *Subsistence Waterfowl Harvest Survey, Yukon-Kuskokwim Delta, 1995–1999.* Anchorage, AK: US Fish and Wildlife Service and Yukon Delta National Wildlife Refuge.

Wetlands International. 2012. *Waterbird Population Estimates.* 5th ed. Summary Report. Wageningen, The Netherlands: Wetlands International.

Wolfe, R. J., and A. W. Paige. 1995. *The Subsistence Harvest of Black Brant, Emperor Geese, and Eider Ducks in Alaska.* Division of Subsistence, Paper 234. Juneau: Alaska Department of Fish and Game.

Species References

Emperor Goose

Byrd, G. V., J. C. Williams, and A. Durand. 1992. *Observations of Emperor Geese in the Aleutian Islands during Winter of 1991–1992.* Adak, AK: US Fish and Wildlife Service.

Cottam, C., and P. Knappen. 1939. Food of some uncommon North American birds. *Auk* 56: 138–169.

Dau, C. P., and E. J. Mallek. 2009. *Aerial Survey of Emperor Geese and Other Waterbirds in Southeastern Alaska, Spring 2009.* Anchorage, AK: US Fish and Wildlife Service.

Dau, C. P., K. S. Bollinger, E. J. Mallek, and R. A. Stehn. 2006. *Monitoring the Emperor Goose Population by Aerial Counts and Fall Age Ratio, Fall 2006.* Anchorage, AK: US Fish and Wildlife Service.

Eisenhauer, D. I. 1976. Biology and behavior of the emperor goose (*Anser canagicus* Sewastianov) in Alaska. MS thesis. Purdue University, Lafayette, IN.

Eisenhauer, D. I., and C. M. Kirkpatrick. 1977. *Ecology of the Emperor Goose in Alaska.* Wildlife Monographs 57. Washington, DC: The Wildlife Society.

Eldridge, and F. J. Broerman. 2008. Body mass of prefledging emperor geese *Chen canagica*: large-scale effects of interspecific densities and food availability. *Ibis* 150: 527–540.

Frazer, D. A., and C. M. Kirkpatrick. 1979. Parental and brood behaviour of emperor geese in Alaska. *Wildfowl* 30: 75–85.

Hupp, J. W., J. A. Schmutz, and C. R. Ely. 2006. The prelaying interval of emperor geese on the Yukon-Kuskokwim Delta, Alaska. *Condor* 108: 912–924.

Hupp, J. W., J. A. Schmutz, and C. R. Ely. 2008a. Seasonal survival of radio-marked emperor geese in western Alaska. *Journal of Wildlife Management* 72: 1584–1595.

Hupp, J. W., J. A. Schmutz, and C. R. Ely. 2008b. The annual migration cycle of emperor geese in western Alaska. *Arctic* 61: 23–34.

Hupp, J. W., J. A. Schmutz, C. R. Ely, E. E. Syroechkovskiy, Jr., A. V. Kondratyev, W. D. Eldridge, and E. Lappo. 2007. Moult migration of emperor geese (*Chen canagica*) between Alaska and Russia. *Journal of Avian Biology* 38: 462–470.

Emperor goose, adult swimming

Kistchinski, A. A. 1971. Biological notes on the emperor goose in north-east Siberia. *Wildfowl* 22: 29–34.

Laing, K. K., and D. G. Raveling. 1993. Habitat and food selection by emperor geese goslings. *Condor* 95: 879–888.

Lake, B. C., J. A. Schmutz, M. S. Lindberg, C. R. Ely, W. D. Eldridge, and F. J. Broerman. 2008. Body mass of prefledging emperor geese *Chen canagica*: large-scale effects of inter-specific densities and food availability. *Ibis* 150: 527–540.

Petersen, M. R. 1983. Observations of emperor geese feeding at Nelson Lagoon, Alaska. *Condor* 85: 367–368.

Petersen, M. R. 1990. Nest-site selection by emperor geese and cackling Canada geese. *Wilson Bulletin* 102: 413–426.

Petersen, M. R. 1991. Reproductive ecology of emperor geese. PhD dissertation. University of California, Davis.

Petersen, M. R. 1992a. Reproductive ecology of emperor geese: Annual and individual variation in nesting. *Condor* 94:383–397.

Petersen, M. R. 1992b. Intraspecific variation in egg shape among individual emperor geese. *Journal of Field Ornithology* 63: 344–354.

Petersen, M. R. 1992c. Reproductive ecology of emperor geese: Survival of adult females. *Condor* 94: 398–406.

Petersen, M. R., and R. E. Gill, Jr. 1982. Population status of emperor geese along the north side of the Alaska Peninsula. *Wildfowl* 33: 31–38.

Schmutz, J. A. 1993. Survival and pre-fledging body mass in juvenile emperor geese. *Condor* 95: 222–225.

Schmutz, J. A. 1994. Age, habitat, and tide effects on feeding activity of emperor geese during autumn migration. *Condor* 96: 46–51.

Schmutz, J. A. 2000a. Age-specific breeding in emperor geese. *Wilson Bulletin* 112: 261–263.

Schmutz, J. A. 2000b. Survival and brood rearing ecology of emperor geese. PhD dissertation. University of Alaska–Fairbanks.

Schmutz, J. A. 2001. Selection of habitats by emperor geese during brood rearing. *Waterbirds* 24: 394–401.

Schmutz, J. A., and A. V. Kondratyev. 1995. Evidence of emperor geese breeding in Russia and staging in Alaska. *Auk* 112: 1037–1038.

Schmutz, J. A., and K. K. Laing. 2002. Variation in foraging behavior and body mass in broods of emperor geese (*Chen canagica*): evidence for interspecific density dependence. *Auk* 119: 996–1009.

Schmutz, J. A., B. F. J. Manly, and C. P. Dau. 2001. Effects of gull predation and weather on survival of emperor goose goslings. *Journal of Wildlife Management* 65: 248–257.

Schmutz, J. A., R. F. Rockwell, and M. R. Petersen. 1997. Relative effects of survival and reproduction on the population dynamics of emperor geese. *Journal of Wildlife Management* 61: 191–201.

Schmutz, J. A., S. E. Cantor, and M. R. Petersen. 1994. Seasonal and annual survival of emperor geese. *Journal of Wildlife Management* 58: 525–535.

Thompson, S. C., and D. G. Raveling. 1987. Incubation behavior of emperor geese compared with other geese: interactions of predation, body size, and energetics. *Auk* 104: 707–716.

Watkins, V. S. 2006. Intraspecific nest parasitism in emperor geese (*Chen canagica*): A population genetic analysis. MS thesis. Alaska Pacific University, Anchorage, AK.

Wolfe, R. J., and A. W. Paige. 1995. *The Subsistence Harvest of Black Brant, Emperor Geese, and Eider Ducks in Alaska.* Division of Subsistence Paper 234. Juneau: Alaska Department of Fish and Game.

Greater White-fronted Goose

Ackerman, J. T., J. Y. Takekawa, D. L. Orthmeyer, J. P. Fleskes, J. L. Yee, and K. L. Kruse. 2006. Spatial use by wintering greater white-fronted geese relative to a decade of habitat change in California's Central Valley. *Journal of Wildlife Management* 70: 965–976.

Alisauskas, R. T., and M. S. Lindberg. 2002. Effects of neckbands on survival and fidelity of white-fronted and Canada geese captured as non-breeding adults. *Journal of Applied Statistics* 29: 521–537.

Anderson, J. T., and D. A. Haukos. 2003. Breeding ground affiliation and movements of greater white-fronted geese staging in northwestern Texas. *Southwestern Naturalist* 48: 365–372.

Bauer, R. D. 1979. Historical and status report of the tule white-fronted goose. Pp. 44–55 in: R. L. Jarvis and J. C. Bartonek, eds. *Management and Biology of Pacific Flyway Geese.* Corvallis, OR: OSU Book Stores.

Boyd, H. 1954. White-fronted goose statistics. *Wildfowl Trust Annual Report* 6: 73–79.

Boyd, H. 1958. The survival of Greenland white-fronted geese ringed in Greenland. *Dansk Ornithologisk Forenings Tidsskrift* 52: 1–8.

Campbell, B. [H.], and E. Goodwin. 1985. Breeding age of the tule white-fronted goose. *Journal of Field Ornithology* 56: 286.

"C. S. R." 1977. White-fronted goose. Pp. 403–409 in: S. Cramp and E. L. Simmons, eds. *Handbook of the Birds of Europe, the Middle East, and North Africa: The Birds of the Western Palearctic, Vol. 1. Ostriches to Ducks.* Oxford, UK: Oxford University Press.

Delacour, J., and S. D. Ripley. 1975. Description of a new subspecies of the white-fronted goose, *Anser albifrons. American Museum Novitates* 2565. New York: American Museum of Natural History.

Densmore, R. V., C. R. Ely, K. S. Bollinger, S. Kratzer, M. S. Udevitz, D. J. Fehringer, and T. C. Rothe. 2006. Nesting habitat of the tule greater white-fronted goose *Anser albifrons elgasi*. *Wildfowl* 56: 37–51.

Deuel, B. E., and J. Y. Takekawa. 2008. Tule greater white-fronted goose (*Anser albifrons elgasi*). *Studies of Western Birds* 1: 74–78.

Dunn, J. L. First impressions and other thoughts about tule geese (*Anser albifrons elgasi*). *California Valley Bird Club Bulletin* 8:1–7.

Dzubin, A., H. W. Miller, and G. V. Schildman. 1964. White-fronts. Pp. 135–143 in: J. P. Linduska, ed. *Waterfowl Tomorrow*. Washington, DC: US Government Printing Office.

Elgas, R. 1970. Breeding populations of tule white-fronted geese in northwestern Canada. *Wilson Bulletin*, 82: 420–426.

Ely, C. R. 1989. Extra-pair copulation in the greater white-fronted goose. *Condor* 91: 990–991.

Ely, C. R. 1992. Time allocation by greater white-fronted geese: influence of diet, energy reserves and predation. *Condor* 94: 857–870.

Ely, C. R. 1993. Family stability in greater white-fronted geese. *Auk* 110: 425–435.

Ely, C. R., A. D. Fox, R. T. Alisauskas, A. Andreev, R. G. Bromley, A. G. Degtyarev, B. Ebbinge, E. N. Gurtovaya, R. Kerbes, A. V. Kondratyev, I. Kostin, A. V. Krechmar, K. E. Konstantin, E. Litvin, Y. Miyabayashi, J. H. Mooij, R. M. Oates, D. L. Orthmeyer, Y. Sabano, S. G. Simpson, D. V. Solovieva, M. A. Spindler, Y. V. Syroechkovsky, J. Y. Takekawa, and A. Walsh. 2005. Circumpolar variation in morphological characteristics of greater white-fronted geese *Anser albifrons*. *Bird Study* 52: 104–119.

Ely, C. R., and D. G. Raveling. 1984. Breeding biology of Pacific white-fronted geese. *Journal of Wildlife Management* 48: 823–837.

Ely, C. R., and D. G. Raveling. 1989. Body composition and weight dynamics of wintering greater white-fronted geese. *Journal of Wildlife Management* 53:80–87.

Ely, C. R., and D. G. Raveling. 2011. Seasonal variation in nutritional characteristics of the diet of greater white-fronted geese. *Journal of Wildlife Management* 75: 78–91.

Ely, C. R., and J. A. Schmutz. 1999. *Characteristics of Mid-Continent Greater White-fronted Geese from Interior Alaska: Distribution, Migration Ecology, and Survival*. Central Flyway Technical Committee and Region 7, US Fish and Wildlife Service, Lawton, OK.

Ely, C. R., and J. Y. Takekawa. 1996. Geographic variation in migratory behavior of greater white-fronted geese (*Anser albifrons*). *Auk* 113: 889–901.

Ely, C. R., D. J. Nieman, R. T. Alisauskas, J. A. Schmutz, J. E. Hines, and D. Caswell. 2012. Geographic variation in migration chronology and winter distribution of midcontinent greater white-fronted geese. *Journal of Wildlife Management* 77: 1182–1191.

Fencker, H. 1950. The Greenland white-fronted goose and its breeding-biology. *Dansk Ornithologisk Forenings Tidsskrift*, 44: 61–65.

Hines, J. E., M. O. Wiebe Robertson, M. F. Kay, and S. E. Westover. 2006. Aerial surveys of greater white-fronted geese, Canada geese, and tundra swans on the mainland of the Inuvialuit Settlement Region, western Canadian Arctic, 1989–1993. Pp. 27–43 in: J. E. Hines and M. O. Wiebe Robertson, eds. *Surveys of Geese and Swans in the Inuvialuit Settlement Region, Western Canadian Arctic, 1989–1993*. Occasional Paper 112. Ottawa, ON: Canadian Wildlife Service.

Krapu, G. L., K. J. Reinecke, D. G. Jorde, and S. G. Simpson. 1995. Spring-staging ecology of midcontinent greater white-fronted geese. *Journal of Wildlife Management* 59: 736–746.

Krogman, B. D. 1979. A systematic study of *Anser albifrons* in California. Pp. 22–43 in: R. L. Jarvis and J. L. Bartonek, eds. *Management and Biology of Pacific Flyway Geese*. Corvallis, OR: OSU Book Stores.

Leslie, J. C., and R. H. Chabreck. 1984. Winter habitat preference of white-fronted geese in Louisiana. *Transactions of the North American Wildlife and Natural Resources Conference* 49: 519–526.

Miller, H., A. Dzubin, and I. T. Sweet. 1968. Distribution and mortality of Saskatchewan-banded white-fronted geese. *North American Wildlife and Natural Resources Conference Transactions*, 33: 101–119.

Miller, H., and A. Dzubin. 1965. Regrouping of family members of the white-fronted goose *Anser albifrons* after individual release. *Bird Banding* 36: 184–191.

Mooij, J. H., and C. Zockler. 2000. Reflections on the systematics, distribution, and status of *Anser albifrons*. *Casarca* 6: 92–107.

Orthmeyer, D. L., J. Y. Takekawa, C. R. Ely, M. L. Wege, and W. E. Newton. 1995. Morphological differences in Pacific coast populations of greater white-fronted geese. *Condor* 97: 123–132.

Owen, M. 1972. Some factors affecting food intake and selection in white-fronted geese. *Journal of Animal Ecology* 41: 79–92.

Pearse, A. T., R. T. Alisauskas, G. L. Krapu, and R. R. Cox, Jr. 2011. Changes in nutrient dynamics of midcontinent greater white-fronted geese during spring migration. *Journal of Wildlife Management* 75: 1716–1723.

Schmutz, J. A., and C. R. Ely. 1999. Survival of greater white-fronted geese: Effects of year, season, sex, and body condition. *Journal of Wildlife Management* 63: 1239–1249.

Spaans, B., W. van der Veer, and B. S. Ebbinge. 1999. Cost of incubation in a greater white-fronted goose. *Waterbirds* 22: 151–155.

Spindler, M. A., and M. R. Hans. 2005. *Nesting Biology and Local Movements of Female Greater White-fronted Geese in West-Central Alaska.* Final report 05-01. Koyukuk/Nowitna National Wildlife Refuge, US Fish and Wildlife Service, Galena, AK.

Swarth, H. S., and H. C. Bryant. 1917. A study of the races of the white-fronted goose (*Anser albifrons*) occurring in California. *University of California Publications in Zoology* 17: 209–222.

Timm, D. E., and C. P. Dau. 1979. Productivity, mortality, distribution, and population status of Pacific Flyway white-fronted geese. Pp. 280–298 in: R. L. Jarvis and J. C. Bartonek, eds. *Management and Biology of Pacific Flyway Geese.* Corvallis, OR: OSU Book Stores.

Timm, D. E., M. L. Wege, and D. S. Gilmer. 1982. Current status and management challenges for tule white-fronted geese. *Transactions of the North American Wildlife and Natural Resources Conference* 47: 453–463.

Todd, W. E. C. 1950. Nomenclature of the white-fronted goose. *Condor* 52: 63–68.

Warren, S. M., A. D. Fox, A. Walsh, and P. O'Sullivan. 1992. Age of first pairing and breeding among Greenland white-fronted geese. *Condor* 94: 791–793.

Wilbur, S. R. 1966. The tule white-fronted goose (*Anser albifrons gambeli*) in the Sacramento Valley, California. *Proceedings of the Annual Meeting of the California-Nevada Section of The Wildlife Society* 2: 85–92.

Snow Goose and Ross's Goose

Abraham, K. F., and R. L. Jefferies. 1997. High goose populations: causes, impacts and implications. Pp. 12–82 in: B. D. J. Batt, ed. *Arctic Ecosystems in Peril: Report of the Arctic Goose Habitat Working Group.* Arctic Goose Joint Venture Special Publication. Washington, DC: US Fish and Wildlife Service, and Ottawa, ON: Canadian Wildlife Service.

Abraham, K. F., J. O. Leafloor, and H. G. Lumsden. 1999. Establishment and growth of the lesser snow goose, *Chen caerulescens*, nesting colony on Akimiski Island, James Bay, Northwest Territories. *Canadian Field-Naturalist* 113: 245–250.

Abraham, K. F., P. Mineau, and F. Cooke. 1977. Unusual predators of snow goose eggs. *Canadian Field-Naturalist* 91: 317–318.

Abraham, K. F., P. Mineau, and F. Cooke. 1981. Remating of a lesser snow goose. *Wilson Bulletin* 93: 557–559.

Abraham, K. F., R. L. Jefferies, and R. T. Alisauskas. 2005. The dynamics of landscape change and snow geese in mid-continent North America. *Global Change Biology* 11: 841–855.

Abraham, K. F., R. L. Jefferies, R. T. Alisauskas, and R. F. Rockwell. 2012. Northern wetland ecosystems and their response to high densities of lesser snow geese and Ross's geese. Pp. 9–45 in: J. O. Leafloor, T. J. Moser, and B. D. J.

Batt, editors. *Evaluation of Special Management Measures for Midcontinent Lesser Snow Geese and Ross's Geese. Arctic Goose Joint Venture Special Publication.* Washington, DC: US Fish and Wildlife Service, and Ottawa, ON: Canadian Wildlife Service.

Abraham, K. F., R. L. Jefferies, R. F. Rockwell, and C. D. MacInnes. 1996. Why are there so many white geese in North America? *International Waterfowl Symposium* 7: 79–92.

Alisauskas, R. T. 1998. Winter range expansion and relationships between landscape and morphometrics of midcontinent lesser snow geese. *Auk* 115: 851–862.

Alisauskas, R. T. 2002. Arctic climate, spring nutrition, and recruitment in midcontinent lesser snow geese. *Journal of Wildlife Management* 66: 181–193.

Alisauskas, R. T., and C. D. Ankney. 1992a. The cost of egg laying and its relationship to nutrient reserves in waterfowl. Pp. 30–61 in: B. D. J. Batt et al., eds. *Ecology and Management of Breeding Waterfowl.* Minneapolis: University of Minnesota Press.

Alisauskas, R. T., and C. D. Ankney. 1992b. Spring habitat use and diets of midcontinent adult lesser snow geese. *Journal of Wildlife Management* 56: 43–54.

Alisauskas, R. T., and H. Boyd. 1994. Previously unrecorded colonies of Ross's and lesser snow geese in the Queen Maude Gulf Bird Sanctuary. *Arctic* 47: 69–73.

Alisauskas, R. T., and M. S. Lindberg. 2002. Effects of neckbands on survival and fidelity of white-fronted and Canada geese captured as non-breeding adults. *Journal of Applied Statistics* 29: 521–537.

Alisauskas, R. T., C. D. Ankney, and E. E. Klaas. 1988. Winter diets and nutrition of midcontinent lesser snow geese. *Journal of Wildlife Management* 50: 403–414.

Alisauskas, R. T., J. O. Leafloor, and D. K. Kellett. 2012. Population status of midcontinent lesser snow geese and Ross's geese following special conservation measures. Pp. 132–177 in: J. O. Leafloor, T. J. Moser, and B. D. J. Batt, eds. *Evaluation of Special Management Measures for Midcontinent Lesser Snow Geese and Ross's Geese. Arctic Goose Joint Venture Special Publication.* Washington, DC: US Fish and Wildlife Service, and Ottawa, ON: Canadian Wildlife Service.

Alisauskas, R. T., J. W. Charlwood, and D. K. Kellett. 2006. Vegetation correlates of the history and density of nesting by Ross's geese and lesser snow geese at Karrak Lake, Nunavut. *Arctic* 59: 201–210.

Alisauskas, R. T., K. L. Drake, and J. D. Nichols. 2009. Filling a void: abundance estimation of North American populations of arctic geese using hunter recoveries. *Environmental and Ecological Statistics* 3: 463–489.

Alisauskas, R. T., R. F. Rockwell, K. W. Dufour, E. G. Cooch, G. Zimmerman, K. L. Drake, J. O. Leafloor, T. J. Moser, and E. T. Reed. 2011. *Harvest, Survival, and Abundance of Mid-Continent Lesser Snow Geese Relative to Population Reduction Efforts.* Wildlife Monographs 179. Bethesda, MD: The Wildlife Society.

Alisauskas, R. T, S. M. Slattery, D. K. Kellett, D. Stern, and K. D. Warner. 1998. *Spatial and Temporal Dynamics of Ross's and Snow Goose Colonies in Queen Maud Gulf Bird Sanctuary, 1996–1998: Progress Report on Numbers of Geese and Colonies, September 1998.* Saskatoon, SK: Canadian Wildlife Service.

Ankney, C. D. 1980. Egg weight, survival, and growth of lesser snow goose goslings. *Journal of Wildlife Management* 44: 174–182.

Ankney, C. D. 1982. Annual cycle of body weight in lesser snow geese. *Wildlife Society Bulletin* 10: 60–64.

Ankney, C. D. 1996. An embarrassment of riches: Too many geese. *Journal of Wildlife Management* 60: 217–223.

Armstrong, W. T., K. M. Meeres, R. H. Kerbes, W. S. Boyd, J. G. Silveira, J. P. Taylor, and B. Turner. 1999. Routes and timing of migration of lesser snow geese from the western Canadian Arctic and Wrangel Island, Russia, 1987–1992. Pp. 75–88 in: R. H. Kerbes, K. M. Meeres, and J. E. Hines, eds. *Distribution, Survival, and Numbers of Lesser Snow Geese of the Western Canadian Arctic and Wrangel Island, Russia.* Occasional Paper 98. Ottawa, ON: Canadian Wildlife Service.

Aubin, A. E., E. H. Dunn, and C. D. MacInnes. 1986. Growth of lesser snow geese on Arctic breeding grounds. *Condor* 88: 365–370.

Baranyuk, V. V. 1999. Breeding of the lesser snow goose in limited nesting area. *Casarca* 5: 161–176.

Baranyuk, V. V. 2000. Weather surprises and breeding of the snow geese on Wrangel Island in 2000. *Cascara* 6: 359–364.

Baranyuk, V. V., and J. Y. Takekawa. 1998. Migration of the snow geese in Chukotka and St. Lawrence Island. *Cascara* 4: 169–187.

Baranyuk, V. V., J. E. Hines, and E. V. Syroechkovsky. 1999. Mineral staining of facial plumage as an indicator of the wintering ground affinities of Wrangel Island lesser snow geese. Pp. 111–114 in: R. H. Kerbes, K. M. Meeres, and J. E. Hines, eds. *Distribution, Survival, and Numbers of Lesser Snow Geese of the Western Canadian Arctic and Wrangel Island, Russia.* Occasional Paper 98. Ottawa, ON: Canadian Wildlife Service.

Batt, B. D. J. 1998. *Snow Geese: Grandeur and Calamity on an Arctic Landscape.* Memphis, TN: Ducks Unlimited.

Batt, B. D. J., ed. 1997. *Arctic Ecosystems in Peril: Report of the Arctic Habitat Working Group.* Arctic Goose Joint Venture Special Publication. Washington, DC: US Fish and Wildlife Service, and Ottawa, ON: Canadian Wildlife Service.

Béchet A., A. Reed, N. Plante, J.-F. Giroux, and G. Gauthier. 2004. Estimating the size of the greater snow goose population. *Journal of Wildlife Management* 68: 639–649.

Bélanger, L., and J. Bédard. 1994. Foraging ecology of greater snow geese, *Chen caerulescens atlantica* in different *Scirpus* marsh plant communities. *Canadian Field-Naturalist* 108: 271–281.

Blokpoel, H. 1974. *Migration of Lesser Snow and Blue Geese in Spring across Southern Manitoba, Part 1: Distribution, Chronology, Directions, Numbers, Heights, and Speeds.* Report Series 28. Ottawa, ON: Canadian Wildlife Service.

Blokpoel, H., and M. C. Gauthier. 1975. *Migration of Lesser Snow and Blue Geese in Spring across Southern Manitoba, Part 2: Influence of Weather and Prediction of Major Flights.* Report Series 32. Ottawa, ON: Canadian Wildlife Service.

Bolen, E. G., and M. K. Rylander. 1978. Feeding adaptations in the lesser snow goose *Anser caerulescens. Southwestern Naturalist* 23: 158–161.

Bousfield, M. A., and Y. V. Syroechkovsky. 1985. A review of recent Soviet research on the lesser snow geese on Wrangel Island, USSR, *Wildfowl* 36: 13–20.

Boyd, W. S. 2000. A comparison of photo counts versus visual estimates for determining the size of snow goose flocks. *Journal of Field Ornithology* 71: 686–690.

Boyd, W. S., and F. Cooke. 2000. Changes in the wintering distribution of Wrangel Island snow geese. *Wildfowl* 51: 59–66.

Cooch, E. G., D. B. Lank, A. Dzubin, R. F. Rockwell, and F. Cooke. 1991a. Body size variation in lesser snow geese: Environmental plasticity in gosling growth rates. *Ecology* 72: 503–512.

Cooch, E. G., D. B. Lank, R. F. Rockwell, and F. Cooke. 1989. Long-term decline in fecundity in a snow goose population: evidence for density dependence? *Journal of Animal Ecology* 58: 711–726.

Cooch, E. G., D. B. Lank, R. F. Rockwell, and F. Cooke. 1991. Long-term decline in body size in a snow goose population: evidence of environmental degradation? *Journal of Animal Ecology* 60: 483–496.

Cooch, E. G., D. B. Lank, R. F. Rockwell, and F. Cooke. 1999. Body size and age of recruitment in snow geese (*Anser c. caerulescens*). *Bird Study* 46 (Supplement): 112–119.

Cooch, E. G., R. L. Jefferies, R. F. Rockwell, and F. Cooke. 1993. Environmental change and the cost of philopatry: An example in the lesser snow goose. *Oecologia* 93: 128–138.

Cooch, F. G. 1955. Observations on the autumn migration of blue geese. *Wilson Bulletin* 67: 171–174.

Cooch, F. G. 1958. The breeding biology and management of the blue goose *Chen caerulescens.* PhD dissertation, Cornell University, Ithaca, NY.

Cooch, F. G. 1961. Ecological aspects of the blue-snow goose complex. *Auk* 78: 72–79.

Cooch, F. G. 1963. Recent changes in distribution of color phases of *Chen c. caerulescens. International Ornithological Congress* 13: 1182–1194.

Cooch, F. G., and J. A. Beardmore. 1959. Assortative mating and reciprocal differences in the blue-snow goose complex. *Nature* 183: 1833–1834.

Cooch, F. G., G. M. Stirrett, and G. F. Boyer. 1960. Autumn weights of blue geese *Chen caerulescens. Auk* 77: 460–465.

Cooke, F. 1987. Lesser snow goose: a long-term population study. Pp. 407–432 in: F. Cooke and P. A. Buckley, eds. *Avian Genetics: A Population and Ecological Approach*. London: Academic Press.

Cooke, F., and C. M. McNally. 1975. Mate selection and colour preferences in lesser snow geese. *Behaviour* 53: 151–170.

Cooke, F., and D. S. Sulzbach. 1978. Mortality, emigration, and separation of mated snow geese. *Journal of Wildlife Management* 42: 271–280.

Cooke, F., and F. G. Cooch. 1968. The genetics of polymorphism in the goose *Anser caerulescens*. *Evolution* 22: 289–300.

Cooke, F., and K. F. Abraham. 1980. Habitat and locality selection in lesser snow geese: the role of previous experience. *Proceedings of the International Ornithological Congress* 17: 998–1004.

Cooke, F., and R. Harmson. 1983. Does sex ratio vary with egg sequence in lesser snow geese? *Auk* 100: 215–217.

Cooke, F., C. M. Francis, E. G. Cooch, and R. T. Alisauskas. 2000. Impact of hunting on population growth of midcontinent lesser snow geese. Pp. 17–31 in: H. Boyd, ed. *Population Modelling and Management of Snow Geese*. Occasional Paper 102. Ottawa, ON: Canadian Wildlife Service.

Cooke, F., C. S. Findlay, and R. F. Rockwell. 1984. Recruitment and the timing of reproduction in lesser snow geese (*Chen caerulescens caerulescens*). *Auk* 101: 451–458.

Cooke, F., D. T. Parkin, and R. F Rockwell. 1988. Evidence of former allopatry of the two color phases of lesser snow geese (*Chen caerulescens caerulescens*). *Auk* 105: 467–479.

Cooke, F., G. H. Finney, and R. F. Rockwell. 1976. Assortative mating in lesser snow geese (*Anser caerulescens*). *Behavior Genetics* 6: 127–140.

Cooke, F., M. A. Bousfield, and A. Sadura. 1981. Mate change and reproductive success in the lesser snow goose. *Condor* 83: 322–327.

Cooke, F., R. F. Rockwell, and D. E. Lank. 1995. *The Snow Geese of La Pérouse Bay: Natural Selection in the Wild*. Oxford, UK: Oxford University Press.

Davies, J. C., and F. Cooke. 1983. Intraclutch hatch synchronization in the lesser snow goose. *Canadian Journal of Zoology* 61: 1398–1401.

Davis, S. E., E. E. Klaas, and K. J. Koehler. 1989. Diurnal time-activity budgets and habitat use of lesser snow geese *Anser caerulescens* in the middle Missouri River Valley during winter and spring. *Wildfowl* 40: 45–54.

Dufour, K. W., R. T. Alisauskas, R. F. Rockwell, and E. T. Reed. 2012. Temporal variation in survival and productivity of midcontinent lesser snow geese and survival of Ross's geese and its relation to population reduction efforts. Pp. 95–131, in J. O. Leafloor, T. J. Moser, and B. D. J. Batt, editors. *Evaluation of Special Management Measures for Midcontinent Lesser Snow Geese and Ross's Geese*. Arctic Goose Joint Venture Special Publication. Washington, DC: US Fish and Wildlife Service, and Ottawa, ON: Canadian Wildlife Service.

Dunn, P. O., A. D. Afton, M. L. Gloutney, and R. T. Alisauskas. 1999. Forced copulation results in few extra-pair fertilizations in Ross's and lesser snow geese. *Animal Behaviour* 57: 1071–1081.

Francis, C. M., and F. Cooke. 1992. Sexual differences in survival and recovery rates of lesser snow geese. *Journal of Wildlife Management* 56: 287–296.

Francis, C. M., M. H. Richards, F. Cooke, and R. F. Rockwell. 1992a. Long-term changes in survival rates of lesser snow geese. *Ecology* 73: 1346–1362.

Francis, C. M., M. H. Richards, F. Cooke, and R. F. Rockwell. 1992b. Changes in survival rates of lesser snow geese with age and breeding status. *Auk* 109: 731–747.

Gloutney, M. L., R. T. Alisauskas, A. D. Afton, and S. M. Slattery. 2001. Foraging time and dietary intake by breeding Ross's and lesser snow geese. *Oecologia* 127: 78–86.

Gloutney, M. L., R. T. Alisauskas, K. A. Hobson, and A. D. Afton. 1999. Use of supplemental food by breeding Ross's geese and lesser snow geese: evidence for variable anorexia. *Auk* 116: 97–108.

Hanson, H. C., H. G. Lumsden, J. J. Lynch, and H. W. Norton. 1972. *Population Characteristics of Three Mainland Colonies of Blue and Lesser Snow Geese Nesting in the Southern Hudson Bay Region*. Research Report (Wildlife) 92. Toronto: Ontario Ministry of Natural Resources.

Hanson, H. C., P. Queneau, and P. Scott. 1956. *The Geography, Birds, and Mammals of the Perry River Region.* Arctic Institute of North America Special Publication 3. Montreal, QC: Arctic Institute of North America.

Harpole, D. N., D. E. Gawlick, and R. D. Slack. 1994. Differential winter distribution of Ross's geese and snow geese in Texas. *Proceedings of the Annual Association of Fish and Wildlife Agencies* 48: 14–21.

Harvey, J. M. 1971. Factors affecting blue goose nesting success. *Canadian Journal of Zoology* 49: 223–234.

Hines, J. E., V. V. Baranyuk, B. Turner, W. S. Boyd, J. G. Silveira, J. P. Taylor, S. J. Barry, K. M. Meeres, R. H. Kerbes, and W. T. Armstrong. 1999. Autumn and winter distributions of lesser snow geese from the western Canadian Arctic and Wrangel Island, Russia, 1953–1992. Pp. 39–73 in: R. H. Kerbes, K. M. Meeres, and J. E. Hines, eds. *Distribution, Survival, and Numbers of Lesser Snow Geese of the Western Canadian Arctic and Wrangel Island, Russia.* Occasional Paper 98. Ottawa, ON: Canadian Wildlife Service.

Hobaugh, W. C. 1984. Habitat use by snow geese wintering in southeast Texas. *Journal of Wildlife Management* 48: 1085–1096.

Humphries, E. M., J. L. Peters, J. E. Jonsson, R. Stone, A. D. Afton, and K. E. Omland. 2009. Genetic differentiation between sympatric and allopatric wintering populations of snow geese. *Wilson Journal of Ornithology* 121: 730–738.

Hupp, J. W., D. G. Robertson, and A. W. Brackney. 2002. Snow geese. Pp. 71–74 in: D. C. Douglas, P. E. Reynolds, and E. B. Rhode, eds. *Arctic Refuge Coastal Plain Terrestrial Wildlife Research Summaries. Biological Science Report USGS/BRD/BSR-2002-0001.* Reston, VA: US Geological Survey.

Hupp, J. W., R. G. White, J. S. Sedinger, and D. G. Robertson, 1996. Forage digestibility and intake by lesser snow geese: effects of dominance and resource heterogeneity. *Oecologia* 108: 232–240.

Hupp, J. W., A. B. Zacheis, R. M. Anthony, D. G. Robertson, W. P. Erickson, and K. C. Palacios. 2001. Snow cover and snow goose *Anser caerulescens caerulescens* distribution during spring migration. *Wildlife Biology* 7: 65–76.

Jackson, S. L., D. S. Hik, and R. F. Rockwell. 1988. The influence of nesting habitat on reproductive success of the lesser snow goose. *Canadian Journal of Zoology* 66: 1699–1703.

Jeffrey, R., and G. Kaiser. 1979. The snow goose flock of the Fraser and Skagit Deltas. Pp. 266–279 in: R. L. Jarvis and J. C. Bartonek, eds. *Management and Biology of Pacific Flyway Geese.* Corvallis, OR: OSU Book Stores.

Johnson, M. A., and C. D. Ankney, eds. *Direct Control and Alternative Harvest Strategies for North American Light Geese: Report of the Direct Control and Alternative Harvest Measures Working Group.* Arctic Goose Joint Venture Special Publication. Washington, DC: US Fish and Wildlife Service, and Ottawa, ON: Canadian Wildlife Service.

Johnson, M. A., P. I. Padding, M. H. Gendron, E. T. Reed, and D. A. Graber. 2012. Assessment of harvest from conservation actions for reducing midcontinent light geese and recommendations for future monitoring. Pp. 46–94 in: J. O. Leafloor, T. J. Moser, and B. D. J. Batt, eds. *Evaluation of Special Management Measures for Midcontinent Lesser Snow Geese and Ross's Geese.* Arctic Goose Joint Venture Special Publication. Washington, DC: US Fish and Wildlife Service, and Ottawa, ON: Canadian Wildlife Service.

Johnsgard, P. A, 1974. *Song of the North Wind: A Story of the Snow Goose.* New York: Doubleday, Anchor. 150 pp.

Johnsgard, P. A,. 2010. Snow geese of the Great Plains. *Prairie Fire,* February 2010, pp. 12–15. http://www.prairiefirenewspaper.com/2010/02/snow-geese-on-the-great-plains

Johnson, S. R. 1996. Staging and wintering areas of snow geese nesting on Howe Island, Alaska. *Arctic* 49: 86–93.

Johnson, S. R., and D. M. Troy. 1987. Nesting of the Ross's goose and blue-phase snow goose in the Sagavanirktok River Delta, Alaska. *Condor* 89: 665–667.

Jónsson, J. E. 2005. Effects of body size on goose behavior: lesser snow goose and Ross's goose. PhD dissertation, Louisiana State University, Baton Rouge, LA.

Jónsson, J. E., A. D. Afton, and R. T. Alisauskas. 2007. Does body size influence nest attendance? A comparison of Ross's geese (*Chen rossii*) and the larger, sympatric lesser snow geese (*C. caerulescens caerulescens*). *Journal für Ornithologie* 148: 549–555.

Jónsson, J. E., and A. D. Afton. 2008. Lesser snow geese and Ross's geese form mixed flocks during winter but differ in family maintenance and social status. *Wilson Journal of Ornithology* 120: 725–731.

Jónsson, J. E., and A. D. Afton. 2009. Time budgets of snow geese *Chen caerulescens* and Ross's geese *Chen rossii* in mixed flocks: Implications of body size, ambient temperature, and family associations. *Ibis* 151: 134–144.

Kerbes, R. H. 1975. *The Nesting Population of Lesser Snow Geese in the Eastern Arctic: A Photographic Inventory of June 1973*. Report Series 35. Ottawa, ON: Canadian Wildlife Service.

Kerbes, R. H. 1982. Lesser snow geese and their habitat on west Hudson Bay. *Le Naturaliste Canadien* 109: 905–911.

Kerbes, R. H. 1983. Lesser snow goose colonies in the western Canadian Arctic. *Journal of Wildlife Management* 47: 523–526.

Kerbes, R. H. 1994. *Colonies and Numbers of Ross's Geese and Lesser Snow Geese in the Queen Maud Gulf Migratory Bird Sanctuary*. Occasional Paper 81. Ottawa, ON: Canadian Wildlife Service.

Kerbes, R. H, K. M. Meeres, and J. E. Hines, editors. 1999. *Distribution, Survival, and Numbers of Lesser Snow Geese of the Western Canadian Arctic and Wrangel Island, Russia*. Occasional Paper 98. Ottawa, ON: Canadian Wildlife Service.

Kerbes, R. H., K. M. Meeres, and R. T. Alisauskas. 2014. *Surveys of Nesting Lesser Snow Geese and Ross's Geese in Arctic Canada, 2002–2009*. Arctic Goose Joint Venture Special Report. Washington, DC: US Fish and Wildlife Service, and Ottawa, ON: Canadian Wildlife Service.

Kerbes, R. H., K. M. Meeres, R. T. Alisauskas, F. D. Caswell, K. F. Abraham and R. K. Ross. 2006. *Surveys of Nesting Mid-Continent Lesser Snow Geese and Ross's Geese in Eastern and Central Arctic Canada, 1997–1998*. Technical Report Series 447. Saskatoon, SK: Canadian Wildlife Service, Prairie and Northern Region.

Kerbes, R. H., M. R. McLandress, G. E. J. Smith, G. W. Beyersbergen, and B. Godwin. 1983. Ross's goose and lesser snow goose colonies in the central Canadian Arctic. *Canadian Journal of Zoology* 61: 168–173.

Kerbes, R. H., P. M. Kotanen, and R. L. Jefferies. 1990. Destruction of wetland habitats by lesser snow geese: A keystone species on the west coast of Hudson Bay. *Journal of Applied Ecology* 27: 242–258.

Kerbes, R. H., V. V. Baranyuk, and J. E. Hines. 1999. Estimated size of the western Canadian Arctic and Wrangel Island lesser snow goose populations on their breeding and wintering grounds. Pp. 15–25 in: R. H. Kerbes, K. M. Meeres, and J. E. Hines, eds. *Distribution, Survival, and Numbers of Lesser Snow Geese of the Western Canadian Arctic and Wrangel Island, Russia*. Occasional Paper 98. Ottawa, ON: Canadian Wildlife Service.

Kozlik, F. M., A. W. Miller, and W. C. Rienecker. 1959. Color-marking white geese for determining migration routes. *California Fish and Game* 45: 69–82.

Krapu, G. L., D. A. Brandt, and R. R. Cox, Jr. 2005. Do Arctic-nesting geese compete with Sandhill cranes for waste corn in the central Platte River Valley, Nebraska? *Proceedings of the North American Crane Workshop* 9: 185–191.

Kruse, K. L., D. Fronczak, and D. E. Sharp. 2009. *Light Geese in the Central and Mississippi Flyways*. Denver, CO: US Fish and Wildlife Service.

Lank, D. B., E. G. Cooch, R. F. Rockwell, and F. Cooke. 1989a. Environmental and demographic correlates of intraspecific nest parasitism in lesser snow geese *Chen caerulescens*. *Journal of Animal Ecology* 58: 29–44.

Lank, D. B., M. A. Bousfield, F. Cooke, and R. F. Rockwell. 1991. Why do snow geese adopt eggs? *Behavioral Ecology* 2: 181–187.

Lank, D. B., P. Mineau, R. F. Rockwell, and F. Cooke. 1989b. Intraspecific nest parasitism and extra-pair copulation in lesser snow geese. *Animal Behaviour* 37: 74–89.

Lank, D. B., R. F. Rockwell, and F. Cooke. 1990. Frequency-dependent fitness consequences of intraspecific nest parasitism in snow geese. *Evolution* 44: 1436–1453.

Leafloor, J. O., T. J. Moser, and B. D. J. Batt, eds. 2012. *Evaluation of Special Management Measures for Midcontinent Lesser Snow Geese and Ross's Geese. Arctic Goose Joint Venture Special Publication*. Washington, DC: US Fish and Wildlife Service, and Ottawa, ON: Canadian Wildlife Service.

Lepage, D., G. Gauthier, and A. Reed. 1996. Breeding-site infidelity in greater snow geese: a consequence of constraints on laying date? *Canadian Journal of Zoology* 74: 1866–1875.

Lepage, D., G. Gauthier, and S. Menu. 2000. Reproductive consequences of egg-laying decisions in snow geese. *Journal of Animal Ecology* 69: 414–427.

LeShack, C. R., A. D. Afton, and R. T. Alisauskas. 1998. Effects of male removal on female reproductive biology in Ross's and lesser snow geese. *Wilson Bulletin* 110: 56–64.

Lessells, C. M. 1987. Parental investment, brood size, and time budgets: behaviour of lesser snow geese *Anser c. caerulescens* families. *Ardea* 75: 189–203.

Lessells, C. M., F. Cooke, and R. F. Rockwell. 1989. Is there a trade-off between egg weight and clutch size in wild lesser snow geese (*Anser c. caerulescens*)? *Journal of Evolutionary Biology* 2: 457–472.

Lynch, J. J., and J. R. Singleton. 1964. Winter appraisals of annual productivity in geese and other water birds. *Wildfowl Trust Annual Report* 15: 114–126.

MacInnes, C. D., and R. H. Kerbes. 1988. Growth of the snow goose, *Chen caerulescens*, colony at McConnell River, Northwest Territories, 1940–1980. *Canadian Field-Naturalist* 101: 33–39.

MacInnes, C. D., R. K. Misra, and J. P. Prevett. 1989. Differences in growth parameters of Ross's geese and snow geese: evidence from hybrids. *Canadian Journal of Zoology* 67: 286–290.

McCormick, K. J., and B. D. Arner. 1987. Lesser snow geese, *Chen c. caerulescens*, breeding in Pelly-Lower Garry Lakes area, interior Keewatin District, Northwest Territories. *Canadian Field-Naturalist* 101: 605–607.

McCracken, K. G., A. D. Afton, and R. T. Alisauskas. 1997. Nest morphology and body size of Ross's geese and lesser snow geese. *Auk* 114: 610–618.

McLandress, M. R. 1983a. Temporal changes in habitat selection and nest spacing in a colony of Ross's and lesser snow geese. *Auk* 100: 335–343.

McLandress, M. R. 1983b. Sex, age, and species differences in disease mortality of Ross's and lesser snow geese in California: implications for avian cholera research. *California Fish and Game* 69: 196–206.

McLandress, M. R., and I. McLandress. 1979. Blue-phase Ross's geese and other blue-phase geese in western North America. *Auk* 96: 544–550.

McLaren, P. L., and M. A. McLaren. 1982a. Migration and summer distribution of lesser snow geese in interior Keewatin. *Wilson Bulletin* 94: 494–504.

Mellor, A. A., and R. F. Rockwell. 2006. Habitat shifts and parasite loads of lesser snow geese (*Chen caerulescens caerulescens*). *Ecoscience* 13: 497–502.

Mineau, P., and F. Cooke. 1979. Territoriality in snow geese or the protection of parenthood: Ryder's and Inglis's hypothesis reassessed. *Wildfowl* 30: 16–19.

Nelson, H. K. 1952. Hybridization of Canada geese with blue geese in the wild. *Auk* 69: 425–428.

Oates, R. M., M. McWhorter, G. Muehlenhart, and C. Bitler. 1987. Distribution, abundance, and productivity of fall staging lesser snow geese on coastal habitats of northeast Alaska and northwest Canada, 1985. Pp. 349–366 in: G. W. Garner and P. E. Reynolds, eds. *Arctic National Wildlife Refuge Coastal Plain Resource Assessment: 1985 Update Report, Baseline Study of the Fish, Wildlife, and Their Habitats.* Anchorage, AK: US Fish and Wildlife Service.

Olsen, R. E., and A. D. Afton. 2000. Vulnerability of lesser snow geese to hunting with electronic calling devices. *Journal of Wildlife Management* 64: 983–993.

Pearse, A. T., G. L. Krapu, D. A. Brandt, and P. J. Kinzel. 2010. Changes in agriculture and abundance of snow geese affect carrying capacity of sandhill cranes in Nebraska. *Journal of Wildlife Management* 74: 479–488.

Pearse, A. T., R. T. Alisauskas, G. L. Krapu, and R. R. Cox, Jr. 2012. Spring snow goose hunting influences body composition of waterfowl staging in Nebraska. *Journal of Wildlife Management* 76: 1393–1400.

Prevett, J. P., and C. D. MacInnes. 1980. *Family and Other Social Groups in Snow Geese.* Wildlife Monographs 71. Washington, DC: The Wildlife Society.

Prevett, J. P., I. F. Marshall, and V. G. Thomas. 1979. Fall foods of lesser snow geese in the James Bay region. *Journal of Wildlife Management* 43: 736–742.

Prevett, J. P., I. F. Marshall, and V. G. Thomas. 1985. Spring foods of snow and Canada geese at James Bay. *Journal of Wildlife Management* 49: 558–563.

Ratcliffe, L., R. F. Rockwell, and F. Cooke. 1988. Recruitment and maternal age in lesser snow geese *Chen caerulescens*. *Journal of Animal Ecology* 57: 553–563.

Reed, A., R. Benoit, M. Julien, and R. Lalumière. 1996. *Goose use of the Coastal Habitats of Northeastern James Bay.* Occasional Paper 92. Ottawa, ON: Canadian Wildlife Service.

Reinecker, W. C. 1965. A summary of band returns from lesser snow geese *Chen hyperborea* of the Pacific Flyway. *California Fish and Game* 51: 132–146.

Ryder, J. P. 1975. The significance of territory size in colonial nesting geese: An hypothesis. *Wildfowl* 26: 114–116.

Samelius, G., and R. T. Alisauskas. 2001. Deterring arctic fox predation: The role of parental nest attendance by lesser snow geese. *Canadian Journal of Zoology* 79: 861–866.

Schubert, C. A., and F. Cooke. 1993. Egg-laying intervals in the lesser snow goose. *Wilson Bulletin* 105: 414–426.

Shorey, R. I., K. T. Scribner, J. Kanefsky, M. D. Samuel, and S. V. Libants. 2011. Intercontinental gene flow among western arctic populations of lesser snow geese. *Condor* 113: 735–746.

Slattery, S. M. 1994. Neonate reserves, growth and survival of Ross's and lesser snow goose goslings. MS thesis, University of Saskatchewan, Saskatoon, SK.

Slattery, S. M., and R. T. Alisauskas. 1995. Egg characteristics and body reserves of neonate Ross's and lesser snow geese. *Condor* 97: 970–984.

Slattery, S. M., and R. T. Alisauskas. 2007. Distribution and habitat use of Ross's and lesser snow geese during late brood rearing. *Journal of Wildlife Management* 71: 2230–2237.

Soper, J. D. 1930. *The Blue Goose,* Chen caerulescens *(Linnaeus): An Account of Its Breeding Ground, Migration, Eggs, Nests, and General Habits.* Ottawa, ON: Department of the Interior, Northwest Territories and Yukon Branch.

Trauger, D. L., A. Dzubin, and J. P. Ryder. 1971. White geese intermediate between Ross's geese and lesser snow geese. *Auk* 88: 856–875.

Weaver, M. 2005. *Status of Ross's and Lesser Snow Geese Wintering in California, December 2005.* Sacramento: California Department of Fish and Game.

Wiebe Robertson, M. O., and J. E. Hines. 2006. Aerial surveys of lesser snow goose colonies at Anderson River and Kendall Island, Northwest Territories, 1996–2001. Pp. 58–61 in: J. E. Hines and M. O. W. Robertson, eds. *Surveys of Geese and Swans in the Inuvialuit Settlement Region, Western Canadian Arctic, 1989–2001.* Occasional Paper 112. Ottawa, ON: Canadian Wildlife Service.

Williams, C. K., M. D. Samuel, V. V. Baranyuk, E. G. Cooch, and D. Kraege. 2008. Winter fidelity and apparent survival of lesser snow goose populations in the Pacific Flyway. *Journal of Wildlife Management* 72: 159–167.

Williams, T. D. 1994. Adoption in a precocial species, the lesser snow goose: intergenerational conflict, altruism, or a mutually beneficial strategy? *Animal Behaviour* 47: 101–107.

Williams, T. D., E. G. Cooch, R. L. Jefferies, and F. Cooke. 1993. Environmental degradation, food limitation, and reproductive output: juvenile survival in lesser snow geese. *Journal of Animal Ecology* 62: 766–777.

Williams, T. D., M. J. J. E. Loonen, and F. Cooke. 1994. Fitness consequences of parental behavior in relation to offspring number in a precocial species: the lesser snow goose. *Auk* 111: 563–572.

Wypkema, R. C. P., and C. D. Ankney. 1979. Nutrient reserve dynamics of lesser snow geese staging at James Bay, Ontario. *Canadian Journal of Zoology* 57: 213–219.

Greater Snow Goose

(Citations listed here include only those largely or entirely related to *A. c. atlanticus*; previous citations might also relate to this race.)

Bédard, J., A. Nadeau, A. Gauthier, and G. Gauthier. 1986. Effects of spring grazing by greater snow geese on hay production. *Journal of Applied Ecology* 23: 65–75.

Blokpoel, H., J. D. Heyland, J. Burton, and N. Samson. 1975. Observations of the fall migration of greater snow geese across southern Québec. *Canadian Field-Naturalist* 89: 268–277.

Calvert, A. M., and G. Gauthier. 2005. Effects of exceptional conservation measures on survival and seasonal hunting mortality in greater snow geese. *Journal of Applied Ecology* 42: 442–452.

Féret, M., G. Gauthier, A. Béchet, J.-F. Giroux, and K. A. Hobson. 2003. Effect of a spring hunt on nutrient storage by greater snow geese in southern Québec. *Journal of Wildlife Management* 67: 796–807.

Gauthier, G., J. Bédard, and Y. Bédard. 1984a. Comparison of daily energy expenditure of greater snow geese between two habitats. *Canadian Journal of Zoology* 62: 1304–1307.

Gauthier, G., J. Bédard, J. Huot, and Y. Bédard. 1984b. Spring accumulation of fat by greater snow geese in two staging habitats. *Condor* 86: 192–199.

Gauthier, G., R. J. Hughes, A. Reed, J. Beaulieu, and L. Rochefort. 1995. Effect of grazing by greater snow geese on the production of graminoids at an arctic site (Bylot Island, NWT, Canada). *Journal of Ecology* 83: 653–664.

Gauthier, G., R. Pradel, S. Menu, and J.-D. Lebreton. 2001. Seasonal survival of greater snow geese and effect of hunting under dependence in sighting probability. *Ecology* 82: 3105–3119.

Giroux, J.-F., and J. Bédard. 1987. The effects of grazing by greater snow geese on the vegetation of tidal marshes in the St. Lawrence Estuary. *Journal of Applied Ecology* 24: 773–788.

Giroux, J.-F., and J. Bédard. 1988. Age differences in the fall diet of greater snow geese in Québec. *Condor* 90: 731–734.

Giroux, J.-F., Y. Bédard, and J. Bédard. 1984. Habitat use by greater snow geese during the brood-rearing period. *Arctic* 37: 155–160.

Hughes, R. J., A. Reed, and G. Gauthier. 1994. Space and habitat use by greater snow goose broods on Bylot Island, Northwest Territories. *Journal of Wildlife Management* 58: 536–545.

Hughes, R. J., G. Gauthier, and A. Reed. 1994. Summer habitat use and behaviour of greater snow geese *Anser caerulescens*. *Wildfowl* 45: 49–64.

Lemieux, L. 1959. Histoire naturelle et aménagement de la Grande Oie Blanche *Chen hyperborea atlantica*. *Le Naturaliste Canadien* 86: 133–192.

Lemieux, L. 1959. The breeding biology of the greater snow goose on Bylot Island, Northwest Territories. *Canadian Field-Naturalist* 73: 117–128.

Lepage, D., D. N. Nettleship, and A. Reed. 1998. Birds of Bylot Island and adjacent Baffin Island, Northwest Territories, Canada, 1979 to 1997. *Arctic* 51: 125–141.

Lesage, L., and G. Gauthier. 1997. Growth and organ development in greater snow goose goslings. *Auk* 114: 229–241.

Lesage, L., and G. Gauthier. 1998. Effect of hatching date on body and organ development in greater snow goose goslings. *Condor* 100: 316–325.

Mainguy, J., G. Gauthier, J.-F. Giroux, and J. Bêty. 2006. Gosling growth and survival in relation to brood movements in greater snow geese (*Chen caerulescens atlantica*). *Auk* 123: 1077–1089.

Maisonneuve, C., and J. Bédard. 1992. Chronology of autumn migration by greater snow geese. *Journal of Wildlife Management* 56: 55–62.

Maisonneuve, C., and J. Bédard. 1993. Distribution and movements of greater snow geese, *Chen caerulescens atlantica*, during fall staging in the St. Lawrence Estuary, Québec. *Canadian Field-Naturalist* 107: 305–313.

Reed, A., J.-F. Giroux, and G. Gauthier. 1998. Population size, productivity, harvest, and distribution. Pp. 5–31 in: B. D. J. Batt, ed. *The Greater Snow Goose: Report of the Arctic Goose Habitat Working Group. Arctic Goose Joint Venture Special Publication*. Washington, DC: US Fish and Wildlife Service, and Ottawa, ON: Canadian Wildlife Service.

Reed, A., R. J. Hughes, and G. Gauthier. 1995. Incubation behavior and body mass of female greater snow geese. *Condor* 97: 993–1001.

Reed, E. T., and A. M. Calvert, eds. 2007. *Evaluation of the Special Conservation Measures for Greater Snow Geese: Report of the Greater Snow Goose Working Group*. Arctic Goose Joint Venture Special Publication. Canadian Wildlife Service, Québec Region, Sainte-Foy, QC.

Reed, E. T., J. Bêty, J. Mainguy, G. Gauthier, and J.-F. Giroux. 2003. Molt migration in relation to breeding success in greater snow geese. *Arctic* 56: 76–81.

Tremblay, J.-P., G. Gauthier, D. Lepage, and A. Desrochers. 1997. Factors affecting nesting success in greater snow geese: Effects of habitat and association with snowy owls. *Wilson Bulletin* 109: 449–461.

Ross's Goose

(Citations listed here include only those that largely or entirely concern *A. rossi.*)

Alisauskas, R. T. 2001. Species description and biology. Pp. 5–9 in: T. J. Moser, ed. *The Status of Ross's Geese. Arctic Goose Joint Venture Special Publication.* Washington, DC: US Fish and Wildlife Service, and Ottawa, ON: Canadian Wildlife Service.

Alisauskas, R. T., and R. F. Rockwell. 2001. Population dynamics of Ross's geese. Pp. 55–67 in: T. J. Moser, ed. *The Status of Ross's Geese. Arctic Goose Joint Venture Special Publication.* Washington, DC: US Fish and Wildlife Service, Ottawa, ON: Canadian Wildlife Service.

Alisauskas, R. T, K. L. Drake, S. M. Slattery, and D. K. Kellett. 2006. Neckbands, harvest, and survival of Ross's geese from Canada's central Arctic. *Journal of Wildlife Management* 70: 89–100.

Caswell., J. H. 2009. Population biology of Ross's geese at McConnell River, Nunavut. PhD dissertation, University of Saskatchewan, Saskatoon, SK.

Cooke, F., and J. P. Ryder. 1971. The genetics of polymorphism in the Ross's goose (*Anser rossii*). *Evolution* 25: 483–490.

Didiuk, A. B., R. T. Alisauskas, and R. F. Rockwell. 2001. Interactions with arctic and subarctic habitats. Pp. 19–32 in: T. J. Moser, ed. *The Status of Ross's Geese. Arctic Goose Joint Venture Special Publication.* Washington, DC: US Fish and Wildlife Service, and Ottawa, ON: Canadian Wildlife Service.

Drake, K. L., and R. T. Alisauskas. 2004. Breeding dispersal by Ross's geese in the Queen Maud Gulf metapopulation. *Animal Biodiversity and Conservation* 27: 331–341.

Johnsgard, P. A. 2014. What are blue Ross's geese? *Nebraska Bird Review* 82: 81–85.

Kelley, J. R., D. C. Duncan, and D. R. Yparraguirre. 2001. Distribution and abundance. Pp. 11–18 in: T. J. Moser, ed. *The Status of Ross's Geese. Arctic Goose Joint Venture Special Publication.* Washington, DC: US Fish and Wildlife Service, and Ottawa, ON: Canadian Wildlife Service.

Kerbes, R. H. 1994. *Colonies and Numbers of Ross's Geese and Lesser Snow Geese in the Queen Maud Gulf Migratory Bird Sanctuary.* Occasional Paper No. 81. Ottawa, ON: Canadian Wildlife Service.

MacInnes, C. D., and F. C. Cooch. 1963. Additional eastern records of Ross's goose. *Auk* 80: 77–78.

McLandress, M. R. 1979. Status of Ross's geese in California. Pp. 255–265 in: R. L. Jarvis and J. C. Bartonek, eds. *Management and Biology of Pacific Flyway Geese.* Corvallis, OR: OSU Book Stores.

McLandress, M. R. 1983. Winning with warts? A threat posture suggests a function for caruncles in Ross's geese. *Wildfowl* 34: 5–9.

Melinchuk, R., and J. P. Ryder. 1980. The distribution, fall migration routes, and survival of Ross's geese. *Wildfowl* 31: 161–171.

Moser, T. J., ed. *The Status of Ross's Geese. Arctic Goose Joint Venture Special Publication.* Washington, DC: US Fish and Wildlife Service, and Ottawa, ON: Canadian Wildlife Service.

Prevett, J. P., and C. D. MacInnes. 1972. The number of Ross's geese in central North America. *Condor* 74: 431–438.

Ryder, J. P. 1964. A preliminary study of the breeding biology of Ross's goose. *Wildfowl Trust Annual Report* 15: 127–137.

Ryder, J. P. 1967. *The Breeding Biology of Ross's Goose in the Perry River Region, Northwest Territories.* Report Series 3. Ottawa, ON: Canadian Wildlife Service.

Ryder, J. P. 1969a. Nesting colonies of Ross's goose. *Auk* 86: 282–292.

Ryder, J. P. 1969b. Timing and spacing of nests and breeding biology of Ross's goose. PhD dissertation, University of Saskatchewan, Saskatoon, SK.

Ryder, J. P. 1970. A possible factor in the evolution of clutch size in Ross's goose. *Wilson Bulletin* 82: 5–13.

Ryder, J. P. 1972. Biology of nesting Ross's geese. *Ardea* 60: 185–215.

Canada and Cackling Geese

Abraham, K. F., J. O. Leafloor, and D. H. Rusch. 1999. Molt migrant Canada geese in northern Ontario and western James Bay. *Journal of Wildlife Management* 63: 649–655.

Akesson, T. R., and D. G. Raveling. 1982. Behaviors associated with seasonal reproduction and long-term monogamy in Canada geese. *Condor* 84: 188–196.

Aldrich, T. W. 1983. Behavior and energetics of nesting Canada geese. MS thesis, University of California, Davis.

Alisauskas, R. T., and M. S. Lindberg. 2002. Effects of neckbands on survival and fidelity of white-fronted and Canada geese captured as non-breeding adults. *Journal of Applied Statistics* 29: 521–537.

Atlantic Flyway Council. 2008a. *Management Plan for the North Atlantic Population of Canada Geese.* Technical Game Bird Section, Canada Goose Committee, Atlantic Flyway Council, Washington, DC.

Atlantic Flyway Council. 2008b. *A Management Plan for the Atlantic Population of Canada Geese.* Technical Game Bird Section, Canada Goose Committee, Atlantic Flyway Council, Washington, DC.

Atlantic Flyway Council. 2011. *Atlantic Flyway Resident Population Canada Goose Management Plan.* Technical Game Bird Section, Canada Goose Committee, Atlantic Flyway Council, Washington, DC.

Baker, A. J. 1998. Identification of Canada goose stocks using restriction analysis of mitochondrial DNA. Pp. 435–443 in: D. H. Rusch et al., eds. *Biology and Management of Canada Geese. Proceedings of the International Canada Goose Symposium.* Milwaukee, WI: Dembar Educational Research Services.

Balham, R. W. 1954. The behavior of the Canada goose *Branta canadensis* in Manitoba. PhD dissertation, University of Missouri, Columbia, MO.

Ball, I. J., E. L. Bowhay, and C. F. Yocom. 1981. *Ecology and Management of the Western Canada Goose in Washington.* Biological Bulletin 17. Olympia: Washington Department of Game.

Bell, R. Q., and W. D. Klimstra. 1970. Feeding activities of Canada geese in southern Illinois. *Transactions of the Illinois State Academy of Science* 63: 295–304.

Blokpoel, H., and M. C. Gauthier. 1980. Weather and the migration of Canada geese across southwestern Ontario in spring 1975. *Canadian Field-Naturalist* 94: 293–299.

Bruggink, J. G., T. C. Tacha, J. C. Davies, and K. F. Abraham. 1994. *Nesting and Brood-Rearing Ecology of Mississippi Valley Population Canada Geese.* Wildlife Monographs 126. Bethesda, MD: The Wildlife Society.

Cadieux, M-C., G. Gauthier, and R. J. Hughes. 2005. Feeding ecology of Canada geese (*Branta canadensis interior*) in sub-arctic inland tundra during brood-rearing. *Auk* 122: 144–147.

Campbell, B. H. 1990. Factors affecting the nesting success of dusky Canada geese, *Branta canadensis occidentalis*, on the Copper River Delta, Alaska. *Canadian Field-Naturalist* 104: 567–574.

Chabreck, R. H., H. H. Dupuie, and D. J. Belsom. 1974. Establishment of a resident breeding flock of Canada geese in Louisiana. *Proceedings of the Southeastern Association of Game and Fish Commissioners* 28: 442–455.

Chapman, J. A., C. J. Henry, and H. M. Wright. 1969. The status, population dynamics, and harvest of the dusky Canada goose. *Wildlife Monographs* 18. 48 pp.

Cline, M. L., B. D. Dugger, C. R. Paine, J. D. Thompson, R. A. Montgomery, and K. M. Dugger. 2004. Factors influencing nest survival of giant Canada geese in northeastern Illinois. In T. J. Moser et al., eds. *Proceedings of the 2003 International Canada Goose Symposium*, Madison, WI.

Coleman, T. S., and D. A. Boag. 1987. Foraging characteristics of Canada geese on the Nisutlin River Delta, Yukon. *Canadian Journal of Zoology* 65: 2358–2361.

Collias, N. E., and L. R. Jahn. 1959. Social behavior and breeding success in Canada geese (*Branta canadensis*) confined under semi-natural conditions. *Auk* 76: 478–509.

Collins, D. P., and R. E. Trost. 2009. *2009 Pacific Flyway Data Book.* Portland, OR: Division of Migratory Bird Management, US Fish and Wildlife Service.

Coluccy, J. M., D. A. Graber, and R. D. Drobney. 2004. Population modeling for giant Canada geese and implications for management. Pp. 181–186 in: T. J. Moser, et al., eds. *Proceedings of the 2003 International Canada Goose Symposium, Madison, Wisconsin*. Madison, WI: Dembar Educational Research Services.

Combs, D. L., B. Oretgo, and J. E. Kennamer. 1984. Nesting biology of a resident flock of Canada geese. *Proceedings of the Southeastern Association of Fish and Wildlife Agencies* 38: 228–238.

Conover, M. R. 2011. Population growth and movements of Canada geese in New Haven County, Connecticut, during a 25-year period. *Waterbirds* 34:412–421.

Craighead, J. J., and D. S. Stockstad. 1964. Breeding age of Canada geese. *Journal of Wildlife Management* 28: 57–64.

Craven, S. R., and D. H. Rusch. 1983. Winter distribution and affinities of Canada geese marked on Hudson and James Bays. *Journal of Wildlife Management* 47: 307–319.

Craven, S. R., G. A. Bartelt, D. H. Rusch, and R. E. Trost. 1985. *Distribution and Movement of Canada Geese in Response to Management Changes in Central Wisconsin, 1975–1981*. Technical Bulletin 158. Madison: Wisconsin Department of Natural Resources.

Davis, R. A., R. N. Jones, C. D. MacInnes, and A. J. Pakulak. 1985. Molt migration of large Canada geese on the west coast of Hudson Bay. *Wilson Bulletin* 97: 296–305.

Deviche, P., and M. Moore. 2007. Arizona white-cheeked geese: The Canada vs. cackling goose identification challenge. *Arizona Field Naturalists*. http://azfo.org/gallery/challenges/WhiteCheekedGeese.html

Dickson, K. M. 2000. The diversity of Canada geese. Pp. 11–24 in: K. M. Dickson, ed. *Towards Conservation of the Diversity of Canada Geese* (Branta canadensis). Occasional Paper 103. Ottawa, ON: Canadian Wildlife Service.

Didiuk, A. B., and D. H. Rusch. 1998. Movements of Canada goose broods near Cape Churchill, Manitoba. Pp. 79–85 in: D. H. Rusch, et al., eds. *Biology and Management of Canada Geese: Proceedings of the 1998 International Canada Goose Symposium, Milwaukee, Wisconsin*. Madison, WI: Dembar Educational Research Services.

Dunn, E. H., and C. D. MacInnes. 1987. Geographic variation in clutch size and body size of Canada geese. *Journal of Field Ornithology* 58: 355–371.

Eichholz, M. W., and J. S. Sedinger. 2007. Survival and recovery rate of Canada geese staging in interior Alaska. *Journal of Wildlife Management* 71: 36–42.

Ely, C. R., J. M. Pearce, and R. W. Ruess. 2008. Nesting biology of lesser Canada geese, *Branta canadensis parvipes*, along the Tanana River, Alaska. *Canadian Field-Naturalist* 122: 29–33.

Erskine, A. J. 2000. Canada geese in the Maritime Provinces. Pp. 85–89 in: K. M. Dickson, ed. *Towards Conservation of the Diversity of Canada Geese* (Branta canadensis). Occasional Paper 103. Ottawa, ON: Canadian Wildlife Service.

Erskine, A. J., and F. Payne. 1997. Canada geese staging and wintering in Yarmouth County, Nova Scotia. Pp. 14–16 in: A. J. Erskine, ed. *Canada Goose Studies in the Maritime Provinces, 1950–1992*. Occasional Paper 7. Sackville, NB: Canadian Wildlife Service, Atlantic Region.

Fondell, T. F., J. B. Grand, D. A. Miller, and R. M. Anthony. 2006. Renesting by dusky Canada geese on the Copper River Delta, Alaska. *Journal of Wildlife Management* 70: 955–964.

Fondell, T. F., J. B. Grand, D. A. Miller, and R. M. Anthony. 2008. Predators of dusky Canada goose goslings and the effect of transmitters on gosling survival. *Journal of Field Ornithology* 79: 399–407.

Fox, A. D., and C. M. Glahder. 2010. Post-moult distribution and abundance of white-fronted geese and Canada geese in West Greenland in 2007. *Polar Research* 29: 413–420.

Fritzell, P. A., Jr., and D. R. Luukkonen. 2004. Influence of banding distribution on harvest indices of Mississippi Valley Population Canada geese. Pp. 51–59 in: T. J. Moser et al., eds. *Proceedings of the 2003 International Canada Goose Symposium*, Madison, WI.

Gates, R. J., D. F. Caithamer, and T. C. Tacha. 1998. Bioenergetics of Canada geese during breeding and postbreeding in northern Ontario. Pp. 323–335 in: D. H. Rusch, et al., eds. *Biology and Management of Canada Geese: Proceedings of the 1998 International Canada Goose Symposium, Milwaukee, Wisconsin*. Madison, WI: Dembar Educational Research Services.

Gates, R. J., D. F. Caithamer, T. C. Tacha, and C. R. Paine. 1993. The annul molt cycle of *Branta canadensis interior* in relation to nutrient reserve dynamics. *Condor* 95: 680–693.

Gates, R. J., D. F. Caithamer, W. E. Moritz, and T. C. Tacha. 2001. *Bioenergetics and Nutrition of Mississippi Valley Population Canada Geese during Winter and Migration.* Wildlife Monographs 146. Bethesda, MD: The Wildlife Society.

Gleason, J. S., K. F. Abraham, C. D. Ankney, and J. O. Leafloor. 2004. Variation in reproductive performance of Canada geese in the presence and absence of lesser snow geese. Pp. 75–83 in: T. J. Moser, et al., editors. *Proceedings of the 2003 International Canada Goose Symposium*, Madison, WI.

Grand, J. B., T. F. Fondell, D. A. Miller, and R. M. Anthony. 2006. Nest survival in dusky Canada geese (*Branta canadensis occidentalis*): Use of discrete-time models. *Auk* 123: 198–210.

Grieb, J. R. 1970. *The Shortgrass Prairie Canada Goose Population.* Wildlife Monographs 22. Washington, DC: The Wildlife Society.

Griggs, K. M., and J. M. Black. 2004. Assessment of a western Canada goose translocation: landscape use, movement patterns, and population viability. Pp. 231–239 in: T. J. Moser, et al., eds. *Proceedings of the 2003 International Canada Goose Symposium.* Madison, WI: Dembar Educational Research Services.

Hainsworth, F. R. 1987. Precision and dynamics of positioning by Canada geese flying in formation. *Journal of Experimental Biology* 128: 445–462.

Hanson, H. C. 1949. Methods of determining age in Canada geese and other waterfowl. *Journal of Wildlife Management* 13: 177–183.

Hanson, H. C. 1997. *The Giant Canada Goose.* Rev. ed. Urbana: Southern Illinois University Press.

Hanson, H. C. 2006. *The White-cheeked Geese* Branta canadensis, B. maxima, B. "lawrensis," B. hutchinsii, B. leucopareia, *and* B. minima: *Taxonomy, Ecophysiographic Relationships, Biogeography, and Evolutionary Considerations. Volume 1: Eastern Taxa.* Blythe, CA: Avvar Books.

Hanson, H. C., and R. H. Smith. 1950. Canada geese of the Mississippi Flyway with special reference to an Illinois flock. *Illinois Natural History Survey Bulletin* 25: 67–210.

Hanson, W. C., and L. L. Eberhardt. 1971. *A Columbia River Canada Goose Population, 1950–1970.* Wildlife Monographs 28. Washington, DC: The Wildlife Society.

Hanson, W. C., and R. L. Browning. 1959. Nesting studies of Canada geese on the Hanford Reservation, 1953–1956. *Journal of Wildlife Management* 23: 129–137.

Harris, S. W. 1965. Plumage descriptions and age data for Canada goose goslings. *Journal of Wildlife Management* 29: 874–877.

Heinrich, J. W., and S. R. Craven. 1992. The economic impact of Canada geese at the Horicon marsh, Wisconsin. *Wildlife Society Bulletin* 20: 364–371.

Hestbeck, J. B. 1994. Survival of Canada geese banded in winter in the Atlantic Flyway. *Journal of Wildlife Management* 58: 748–756.

Hestbeck, J. B, and M. C. Bateman. 2000. Breeding, migration, and wintering affinities of Canada geese marked in the Atlantic Provinces. Pp. 73–83 in: K. M. Dickson, ed. *Towards Conservation of the Diversity of Canada Geese (*Branta canadensis*).* Occasional Paper 103. Ottawa, ON: Canadian Wildlife Service.

Hestbeck, J. B., and R. A. Malecki. 1989. Estimated survival rates of Canada geese within the Atlantic Flyway. *Journal of Wildlife Management* 53: 91–96.

Higgins, F. K., and L. J. Schoonover. 1969. Aging small Canada geese by neck plumage. *Journal of Wildlife Management* 33: 212–214.

Hindman, L. J., K. M. Dickson, W. F. Harvey, IV, and J. R. Serie. 2004. Atlantic Flyway Canada geese: new perspectives in goose management. Pp. 12–21 in T. J. Moser, et al., eds. *Proceedings of the 2003 International Canada Goose Symposium.* Madison, WI: Dembar Educational Research Services.

Hindman, L. J., R. A. Malecki, and J. R. Serie. 1996. Status and management of Atlantic Population Canada geese. *International Waterfowl Symposium* 7: 109–116.

Hine, R. L., and C. Schoenfeld, eds. 1968. *Canada Goose Management: Current Continental Problems and Programs.* Madison, WI: Dembar Educational Research Services.

Humburg, D. D., F. D. Caswell, D. H. Rusch, and M. Gillespie. 1998. Breeding ground surveys for the Eastern Prairie Population of Canada geese. Pp. 9–19 in: D. H. Rusch, et al., eds. *Biology and Management of Canada Geese: Proceedings of the 1998 International Canada Goose Symposium, Milwaukee, Wisconsin.* Madison, WI: Dembar Educational Research Services.

Humburg, D. D., F. D. Caswell, D. H. Rusch, M. Gillespie, and P. Telander. 2000. Status and trends of the Eastern Prairie Population of Canada geese. Pp. 123–135 in: K. M. Dickson, ed. *Towards Conservation of the Diversity of Canada Geese* (Branta canadensis). Occasional Paper 103. Ottawa, ON: Canadian Wildlife Service.

Humburg, D. D., D. A. Graber, and K. M. Babcock. 1985. Factors affecting autumn and winter distribution of Canada geese. *Transactions of the North American Wildlife and Natural Resources Conference* 50: 525–539.

Holevinski, R. A., R. A. Malecki, and P. D. Curtis. 2006. Can hunting of translocated nuisance Canada geese reduce local conflicts? *Wildlife Society Bulletin* 34: 845–849.

Imber, M. J. 1968. Sex ratios in Canada goose populations. *Journal of Wildlife Management* 32: 905–920.

Jarvis, R. L., and R. G. Bromley. 1998. Managing racially mixed flocks of Canada geese. Pp. 413–423 in: D. H. Rusch, et al., eds. *Biology and Management of Canada Geese: Proceedings of the 1998 International Canada Goose Symposium, Milwaukee, Wisconsin.* Madison, WI: Dembar Educational Research Services.

Johnson, D. H., D. E. Timm, and P. F. Springer. 1979. Morphological characteristics of Canada geese in the Pacific Flyway. Pp. 56–80 in: R. L. Jarvis and J. C. Bartonek, eds. *Management and Biology of Pacific Flyway Geese.* Corvallis, OR: OSU Book Stores.

Jones, R. M., and M. Obbard. 1970. Canada goose killed by Arctic loon and subsequent pairing of its mate. *Auk* 87: 370–371.

Kaminski, R. M., and H. H. Prince. 1977. Nesting habitat of Canada geese in southeastern Michigan. *Wilson Bulletin* 89: 523–531.

Klopman, R. B. 1958. The nesting of the Canada goose at Dog Lake, Manitoba. *Wilson Bulletin* 70: 168–183.

Klopman, R. B. 1962. Sexual behavior in the Canada goose. *Living Bird* 1: 123–129.

Klopman, R. B. 1968. The agonistic behavior of the Canada goose (*Branta canadensis canadensis*), 1: Attack behavior. *Behaviour* 30: 287–319.

Kossack, C. W. 1950. Breeding habits of Canada geese under refuge conditions. *American Midland Naturalist* 43: 627–649.

Krohn, W. B., and E. G. Bizeau. 1980. *The Rocky Mountain Population of the Western Canada Goose: Its Distribution, Habitats, and Management.* Special Scientific Report – Wildlife 229. Washington, DC: US Fish and Wildlife Service.

Leafloor, J. O., and K. F. Abraham. 2000. Procedures for monitoring the Mississippi Valley Population of Canada geese and suggestions for improvement. Pp. 117–122 in: K. M. Dickson, ed. *Towards Conservation of the Diversity of Canada Geese* (Branta canadensis). Occasional Paper 103. Ottawa, ON: Canadian Wildlife Service.

Leafloor, J. O., K. F. Abraham, D. H. Rusch, R. K. Ross, and M. R. J. Hill. 1996a. Status of the southern James Bay population of Canada geese. *International Waterfowl Symposium* 7: 103–108.

Leafloor, J. O., M. R. J. Hill, D. H. Rusch, K. F. Abraham, and R. K. Ross. 2000. Nesting ecology and gosling survival of Canada geese on Akimiski Island, Northwest Territories, Canada. Pp. 109–116 in: K. M. Dickson, ed. *Towards Conservation of the Diversity of Canada Geese* (Branta canadensis). Occasional Paper 103. Ottawa, ON: Canada Wildlife Service.

Lebeda, C. S., and J. T. Ratti. 1983. Reproductive biology of Vancouver Canada geese on Admiralty Island, Alaska. *Journal of Wildlife Management* 47: 297–306.

Lee, F. B., C. H. Schroeder, T. L. Kuck, L. J. Schoonover, M. A. Johnson, H. K. Nelson, and C. A. Beauduy. 1984. *Rearing and Restoring Giant Canada Geese in the Dakotas.* Bismarck: North Dakota Game and Fish Department.

Lefebvre, E. A., and D. G. Raveling. 1967. Distribution of Canada geese in winter as related to heat loss at varying environmental temperatures. *Journal of Wildlife Management* 31: 538–546.

Lieff, B. C. 1973. The summer feeding ecology of blue and Canada geese at the McConnell River, NWT. PhD dissertation, University of Western Ontario, London, ON.

Luukkonen, D. R., H. H. Prince, and R. C. Mykut. 2008. Movements and survival of molt migrant Canada geese from southern Michigan. *Journal of Wildlife Management* 72: 449–462.

MacInnes, C. D., and B. C. Lief. 1968. Individual behavior and composition of a local population of Canada geese. Pp. 93–101 in: R. L. Hine and C. Schoenfeld, eds. *Canada Goose Management*. Madison, WI: Dembar Educational Research Services.

MacInnes, C. D., and E. H. Dunn. 1988. Estimating proportion of an age class nesting in Canada geese. *Journal of Wildlife Management* 52: 421–423.

Maggiulli, N. G., and B. D. Dugger. 2011. Factors associated with dusky Canada goose (*Branta canadensis occidentalis*) nesting and nest success on artificial nest islands of the western Copper River Delta. *Waterbirds* 34: 269–279.

Malecki, R. A., and R. E. Trost. 1990. A breeding ground survey of Atlantic Flyway Canada geese, *Branta canadensis*, in northern Québec. *Canadian Field-Naturalist* 104: 575–578.

Malecki, R. A., B. D. J. Batt, A. Fox, and S. E. Sheaffer. 2000. *Temporal and Geographic Distribution of North Atlantic Population (NAP) Canada Geese: Final Report. Atlantic Flyway Council Technical Section, July 2000.*

Malecki, R. A., B. D. J. Batt, and S. E. Sheaffer. 2001. Spatial and temporal distribution of Atlantic Population Canada geese. *Journal of Wildlife Management* 65: 242–247.

Malecki, R. A., F. D. Caswell, K. M. Babcock, R. A. Bishop, and R. K. Brace. 1980. Major nesting range of the Eastern Prairie Population of Canada geese. *Journal of Wildlife Management* 44: 229–232.

McLandress, M. R., and D. G. Raveling. 1981a. Hyperphagia and social behavior of Canada geese prior to spring migration. *Wilson Bulletin* 93: 310–324.

McLandress, M. R., and D. G. Raveling. 1981b. Changes in diet and body composition of Canada geese before spring migration. *Auk* 98: 65–79.

McWilliams, S. R., and D. G. Raveling. 1998. Habitat use and foraging behavior of cackling Canada and Ross's geese during spring: implications for the analysis of ecological determinants of goose social behavior. Pp. 167–178 in: D. H. Rusch, et al., eds. *Biology and Management of Canada Geese: Proceedings of the 1998 International Canada Goose Symposium, Madison, Wisconsin.* Madison, WI: Dembar Educational Research Services.

Miller, A. W., and B. D. Collins. 1953. A nesting study of Canada geese on Tule Lake and Lower Klamath National Wildlife Refuges, Siskiyou County, California. *California Fish and Game* 39: 385–396.

Moser, T. J., and D. F. Caswell. 2004. Long-term indices of Canada goose status and management. Pp. 123–129 in: T. J. Moser, et al., eds. *Proceedings of the 2003 International Canada Goose Symposium.* Madison, WI: Dembar Educational Research Services.

Moser T. J., R. D. Lien, K. C. VerCauteren, K. F. Abraham, D. E. Andersen, J. G. Bruggink, J. M. Coluccy, D. A. Graber, J. O. Leafloor, D. R. Luukkonen, and R. E. Trost, eds. 2004. *Proceedings of the 2003 International Canada Goose Symposium, Madison, Wisconsin.* Madison, WI: Dembar Educational Research Services.

Powell, L. A., M. P. Vrtiska, and N. Lyman. 2004. Survival rates and recovery distributions of Canada geese banded in Nebraska. Pp. 60–65 in: T. J. Moser, et al., eds. *Proceedings of the 2003 International Canada Goose Symposium, Madison, Wisconsin.* Madison, WI: Dembar Educational Research Services.

Quinn, T. W., G. F. Shields, and A. C. Wilson. 1991. Affinities of the Hawaiian goose, based on two types of mitochondrial DNA. *Auk* 108: 585–593.

Ratti, J. T., and D. E. Timm. 1979. Migratory behavior of Vancouver Canada geese: recovery rate bias. Pp. 208–212 in: R. L. Jarvis and J. C. Bartonek, eds. *Management and Biology of Pacific Flyway Geese.* Corvallis, OR: OSU Book Stores.

Ratti, J. T., D. E. Timm, and F. C. Robards. 1977. Weights and measurements of Vancouver Canada geese. *Bird-Banding* 48: 354–357.

Raveling, D. G. 1968a. Can counts of group sizes of Canada geese reveal population structure? Pp. 87–91 in Hine and Schoenfeld, eds. *Canada Goose Management: Current Continental Problems and Programs.* Madison, WI: Dembar Educational Research Services.

Raveling, D. G. 1968b. Weights of *Branta canadensis interior* during winter. *Journal of Wildlife Management* 32: 412–414.

Raveling, D. G. 1969a. Social classes of Canada geese in winter. *Journal of Wildlife Management* 33: 304–318.

Raveling, D. G. 1969b. Preflight and flight behavior of Canada geese. *Auk* 86: 671–681.

Raveling, D. G. 1970. Dominance relationships and agonistic behavior of Canada geese in winter. *Behaviour* 37: 291–319.

Raveling, D. G. 1976. Migration reversal: a regular phenomenon of Canada geese. *Science* 193: 153–154.

Raveling, D. G. 1978. Dynamics of distribution of Canada geese in winter. *Transactions of the North American Wildlife and Natural Resources Conference* 43: 206–225.

Raveling, D. G. 1979a. The annual cycle of body composition of Canada geese with special reference to control of reproduction. *Auk* 96: 234–252.

Raveling, D. G. 1979b. The annual energy cycle in the cackling Canada goose. Pp. 81–93 in: R. L. Jarvis and J. C. Bartonek, eds. *Management and Biology of Pacific Flyway Geese.* Corvallis, OR: OSU Book Stores.

Raveling, D. G. 1981. Survival, experience, and age in relation to breeding success of Canada geese. *Journal of Wildlife Management* 45: 817–829.

Raveling, D. G. 1988. Mate retention in giant Canada geese. *Canadian Journal of Zoology* 66: 2766–2768.

Raveling, D. G., and H. G. Lumsden. 1977. *Nesting Ecology of Canada Geese in the Hudson Bay Lowlands of Ontario: Evolution and Population Regulation.* Fish and Wildlife Research Report 98. Toronto: Ontario Ministry of Natural Resources.

Reed, A., R. Benoit, M. Julien, and R. Lalumière. 1996. *Goose Use of the Coastal Habitats of Northeastern James Bay.* Occasional Paper 92. Ottawa, ON: Canadian Wildlife Service.

Reese, K. P., J. A. Kadlec, and L. M. Smith. 1987. Characteristics of islands selected by nesting Canada geese, *Branta canadensis. Canadian Field-Naturalist* 101: 539–542.

Reiter, M. E., and D. E. Andersen. 2011. Arctic foxes, lemmings, and Canada goose nest survival at Cape Churchill, Manitoba. *Wilson Journal of Ornithology* 123: 266–276.

Rienecker, W. C. 1985. Temporal distribution of breeding and non-breeding Canada geese from northeastern California. *California Fish and Game* 71: 196–209.

Rusch, D. H., J. C. Wood, and G. G. Zenner. 1996. The dilemma of giant Canada goose management. *International Waterfowl Symposium* 7: 72–78.

Rusch, D. H., M. D. Samuel, D. D. Humburg, and B. D. Sullivan, eds. 1998. *Biology and Management of Canada Geese: Proceedings of the 1998 International Canada goose Symposium, Milwaukee, Wisconsin.* Madison, WI: Dembar Educational Research Services.

Rusch, D. H., R. A. Malecki, and R. E. Trost. 1995. Canada geese in North America. Pp. 26–28 in: E. T. LaRoe, G. S. Farris, C. E. Puckett, P. D. Doran, and M. J. Mac, eds. *Our Living Resources: A Report to the Nation on the Distribution, Abundance, and Health of US Plants, Animals, and Ecosystems.* Washington, DC: National Biological Service.

Rusch, D. H., S. R. Craven, R. E. Trost, J. R. Cary, R. L. Drieslein, J. W. Ellis, and J. Wetzel. 1985. Evaluation of efforts to redistribute Canada geese. *Transactions of the North American Wildlife and Natural Resources Conference* 50: 506–524.

Samuel, M. D., D. H. Rusch, K. F. Abraham, M. M. Gillespie, J. P. Prevett, and G. W. Swenson. 1991. Fall and winter distribution of Canada geese in the Mississippi Flyway. *Journal of Wildlife Management* 55: 449–456.

Scribner, K. T., R. A. Malecki, B. D. J. Batt, R. L. Inman, S. Libants, and H. H. Prince. 2003a. Identification of source population for Greenland Canada geese: genetic assessment of a recent colonization. *Condor* 105: 771–782.

Scribner, K. T., S. L. Talbot, J. M. Pearce, B. J. Pierson, K. S. Bollinger, and D. V. Derksen. 2003b. Phylogeography of Canada geese (*Branta canadensis*) in western North America. *Auk* 120: 889–907.

Sedinger, J. S. 1986. Growth and development of Canada goose goslings. *Condor* 88: 169–180.

Sheaffer, S. E., and R. A. Malecki. 1998. Status of Atlantic Flyway resident nesting Canada geese. Pp. 29–34 in: D. H. Rusch, et al., eds. *Biology and Management of Canada Geese: Proceedings of the 1998 International Canada Goose Symposium, Milwaukee, Wisconsin.* Madison, WI: Dembar Educational Research Services.

Sheaffer, S. E., D. H. Rusch, D. D. Humburg, J. S. Lawrence, G. G. Zenner, M. M. Gillespie, F. D. Caswell, S. Wilds, and S. C. Yaich. 2004. *Survival, Movements, and Harvest of Eastern Prairie Population Canada Geese.* Wildlife Monographs 158. The Wildlife Society, Bethesda, MD.

Sheaffer, S. E., R. A. Malecki, B. L. Swift, J. Dunn, and K. Scribner. 2008. Management implications of molt migration by the Atlantic Flyway resident population of Canada geese, *Branta canadensis. Canadian Field-Naturalist* 121(3): 313–320.

Sherwood, G. A. 1966. Canada geese of the Seney National Wildlife Refuge. PhD dissertation, Utah State University, Logan, UT.

Shields, G. F., and A. C. Wilson. 1987a. Subspecies of Canada geese (*Branta canadensis*) have distinct mitochondrial DNAs. *Evolution* 76: 2404–2414.

Shields, G. F., and A. C. Wilson. 1987b. Calibration of mitochondrial DNA evolution in geese. *Journal of Molecular Evolution* 24: 212–217.

Shields, G. F., and J. P. Cotter. 1998. Phylogenies of North American geese: The mitochondrial DNA record. In D. H. Rusch, et al., eds. *Biology and Management of Canada Geese. Proceedings of the International Canada Goose Symposium, Milwaukee, Wisconsin.* Madison, WI: Dembar Educational Research Services.

Simpson, S. G., and R. L. Jarvis. 1979. Comparative ecology of several subspecies of Canada geese during winter in western Oregon. Pp. 223–241 in: R. L. Jarvis and J. C. Bartonek, eds. *Management and Biology of Pacific Flyway Geese.* Corvallis, OR: OSU Book Stores.

Sladen, W. J. L., W. A. Lishman, D. H. Ellis, G. G. Shire, and D. L. Rininger. 2002. Teaching migration routes to Canada geese and trumpeter swans using ultralight aircraft, 1990–2001. *Waterbirds* 25 (Special Publication 1): 132–137.

Smith, A. E., and M. R. J. Hill. 1996. Polar bear, *Ursus maritimus,* depredation of Canada goose, *Branta canadensis,* nests. *Canadian Field-Naturalist* 110: 339–340.

Smith, A. E., S. R. Craven, and P. D. Curtis. 1999. *Managing Canada Geese in Urban Environments: A Technical Guide.* [Jack H. Berryman Institute] Publication 16. Cornell Cooperative Extension, Ithaca, NY.

Smith, D. W. 2000. Management of Canada geese in the lower Fraser Valley, southwestern British Columbia. Pp. 151–158 in: K. M. Dickson, ed. *Towards Conservation of the Diversity of Canada Geese (*Branta canadensis*).* Occasional Paper 103. Ottawa, ON: Canadian Wildlife Service.

Steel, P. E., P. D. Dalke, and E. G. Bizeau. 1957. Canada goose production at Gray's Lake, Idaho, 1949–1951. *Journal of Wildlife Management* 21: 38–41.

Sterling, T., and A. Dzubin. 1967. Canada goose molt migrations to the Northwest Territories. *Transactions of the North American Wildlife and Natural Resources Conference* 32: 355–373.

Tacha, T. C., A. Woolf, W. D. Klimstra, and K. F. Abraham. 1991. Migration patterns of the Mississippi Valley population of Canada geese. *Journal of Wildlife Management* 55: 94–102.

Trost, R. E., K. F. Abraham, J. C. Davies, K. E. Bednarik, and H. G. Lumsden. 1998. The distribution of leg-band recoveries from Canada geese banded in the southern James Bay region of Canada. Pp. 249–254 in: D. H. Rusch, et al., eds. *Biology and Management of Canada Geese: Proceedings of the 1998 International Canada Goose Symposium, Milwaukee, Wisconsin.* Madison, WI: Dembar Educational Research Services.

Van Wagner, C. E., and A. J. Baker. 1990. Association between mitochondrial DNA and morphological evolution in Canada geese. *Journal of Molecular Evolution* 31: 373–382.

Vaught, R. W., and G. C. Arthur. 1965. Migration routes and mortality rates of Canada geese banded in the Hudson Bay Lowlands. *Journal of Wildlife Management* 29: 244–252.

Vaught, R. W., and L. M. Kirsch. 1966. *Canada Geese of the Eastern Prairie Population, with Special Reference to the Swan Lake Flock.* Technical Bulletin 3. Jefferson City: Missouri Department of Conservation.

Vermeer, K. 1970a. A study of Canada geese, *Branta canadensis,* nesting on islands in southeastern Alberta. *Canadian Journal of Zoology* 48: 235–240.

Vrtiska, M. P., and N. Lyman. 2004. Wintering Canada geese along the Platte Rivers of Nebraska, 1960–2000. *Great Plains Research* 14: 115–128.

Warhurst, R. A., T. A. Bookhout, and K. E. Bednarik. 1983. Effect of gang brooding on survival of Canada goose goslings. *Journal of Wildlife Management* 47: 1119–1124.

Wege, M. L., and D. G. Raveling. 1983. Factors influencing the timing, distance, and path of migrations of Canada geese. *Wilson Bulletin* 95: 209–221.

Whitford, P. C. 1998. Visual and vocal communication in giant Canada geese. Pp. 375–386 in: D. H. Rusch, et al., eds. *Biology and Management of Canada Geese: Proceedings of the 1998 International Canada Goose Symposium, Milwaukee, Wisconsin*. Madison, WI: Dembar Educational Research Services.

Wood, J. S. 1965. Some associations of behavior to reproductive development in Canada geese. *Journal of Wildlife Management* 29: 237–244.

Yocom, C. F. 1965. Estimated populations of Great Basin Canada geese over their breeding range in western Canada and western United States. *Murrelet* 46: 18–26.

Yocom, C. F. 1972. Weights and measurements of Taverner's and Great Basin Canada geese. *Murrelet*, 53: 33–34.

Zicus, M. C. 1981a. Canada goose brood behavior and survival estimates at Crex Meadows, Wisconsin. *Wilson Bulletin* 93: 207–217.

Zicus, M. C. 1981b. Molt migration of Canada geese from Crex Meadows, Wisconsin. *Journal of Wildlife Management* 45: 54–63.

Zicus, M. C. 1984. Pair separation in Canada geese. *Wilson Bulletin* 96:129–130.

Cackling Goose

(Citations listed here are largely or entirely related to *B. hutchinsii*; see prior citations for broader taxonomic studies that might include *B. hutchinsii*.)

Banks, R. C., C. Cicero, J. L. Dunn, A. W. Kratter, P. C. Rasmussen, J. V. Remsen, J. D. Rising, and D. F. Stotz. 2004. Forty-fifth supplement to the *American Ornithologists' Union Check-list of North American Birds*. Auk 121: 985–995. (Species-level recognition of *Branta hutchinsii*)

Byrd, G. V. 1998. Current breeding status of the Aleutian Canada goose, a recovering endangered species. Pp. 21–28 in: D. H. Rusch, et al., eds. *Biology and Management of Canada Geese: Proceedings of the 1998 International Canada Goose Symposium, Milwaukee, Wisconsin*. Madison, WI: Dembar Educational Research Services.

Bailey, E. P., and J. L. Trapp. 1984. A second wild breeding population of the Aleutian Canada goose. *American Birds* 38: 284–286.

Black, J. M., P. F. Springer, E. T. Nelson, K. M. Griggs, T. D. Taylor, Z. D. Thompson, A. Maguire, and J. Jacobs. 2004. Site selection and foraging behavior of Aleutian Canada geese in a newly colonized spring staging area. Pp. 114–121 in: T. J. Moser, et al., eds. *Proceedings of the 2003 International Canada Goose Symposium, Madison, Wisconsin*. Madison, WI: Dembar Educational Research Services.

Byrd, G. V., and D. W. Wollington. 1983. *Ecology of Aleutian Canada Geese at Buldir Island, Alaska. Special Scientific Report – Wildlife 253*. Washington, DC: US Fish and Wildlife Service.

Byrd, G. V., K. Durbin, F. Lee, T. Rothe, P. Springer, D. Yparraguirre, and F. Zeillermaker. 1991. *Aleutian Canada Goose (Branta canadensis leucopareia) Recovery Plan*. 2nd rev. Anchorage, AK: US Fish and Wildlife Service.

Eichholz, M. W., and J. S. Sedinger. 2006. Staging, migration, and winter distribution of Canada and cackling geese staging in interior Alaska. *Journal of Wildlife Management* 70: 1308–1315.

Ely, C. R. 1998. Survival of cackling Canada geese during brood rearing on the Yukon-Kuskokwim Delta, Alaska [abstract]. Page 100 in: D. H. Rusch, et al., eds. *Biology and Management of Canada Geese: Proceedings of the 1998 International Canada Goose Symposium, Milwaukee, Wisconsin*. Madison, WI: Dembar Educational Research Services.

Fowler, A. C., and C. R. Ely. 1997. Behavior of cackling Canada geese during brood rearing. *Condor* 99: 406–412.

Grieb, J. R. 1970. *The Shortgrass Prairie Canada Goose Population*. Wildlife Monographs 22. Washington, DC: The Wildlife Society.

Hines, J. E., D. L. Dickson, B. C. Turner, M. O. Wiebe, S. J. Barry, T. A. Barry, R. H. Kerbes, D. J. Nieman, M. F. Kay, M. A. Fournier, and R. C. Cotter. 2000. Population status, distribution, and survival of short-grass prairie Canada geese from the Inuvialuit Settlement Region, western Canadian Arctic. Pp. 27–55 in: K. M. Dickson, ed. *Towards Conservation of the Diversity of Canada Geese (Branta canadensis)*. Occasional Paper 103. Ottawa, ON: Canadian Wildlife Service.

Jarvis, R. L., and R. G. Bromley. 2000. Incubation behaviour of Richardson's Canada geese on Victoria Island, Nunavut, Canada. Pp. 59–64 in: K. M. Dickson, ed. *Towards Conservation of the Diversity of Canada Geese (*Branta canadensis*)*. Occasional Paper 103. Ottawa, ON: Canadian Wildlife Service.

MacInnes, C. D. 1962. Nesting of small Canada geese near Eskimo Point, Northwest Territories. *Journal of Wildlife Management*, 26: 247–256.

MacInnes, C. D. 1963. Interactions of local units within the Eastern Arctic population of small Canada geese. PhD dissertation, Cornell University, Ithaca, NY.

MacInnes, C. D. 1966. Population behavior of Eastern Arctic Canada geese. *Journal of Wildlife Management* 30: 536–553.

MacInnes, C. D., and E. H. Dunn. 1988. Estimating proportion of an age class nesting in Canada geese. *Journal of Wildlife Management* 52: 421–423.

MacInnes, C. D., R. A. Davis, R. N. Jones, B. C. Lieff, and A. J. Pakulak. 1974. Reproductive efficiency of McConnell River small Canada geese. *Journal of Wildlife Management* 38: 686–707.

McWilliams, S. R., and D. G. Raveling. 1998. Habitat use and foraging behavior of cackling Canada and Ross's geese during spring: implications for the analysis of ecological determinants of goose social behavior. Pp. 167–178 in: D. H. Rusch, et al., eds. *Biology and Management of Canada Geese: Proceedings of the 1998 International Canada Goose Symposium, Milwaukee, Wisconsin*. Madison, WI: Dembar Educational Research Services.

Mickelson, P. G. 1975. Breeding biology of the cackling goose and associated species on the Yukon-Kuskokwim Delta, Alaska. *Wildlife Monographs* 45.

Mini, A. E., D. C. Bachman, J. Cocke, K. M. Griggs, K. A. Spragens, and J. M. Black. 2011. Recovery of the Aleutian cackling goose *Branta hutchinsii leucopareia:* 10-year review and future prospects. *Wildfowl* 61: 3–29.

Mickelson, P. G. 1973. Breeding biology of cackling geese (*Branta canadensis minima*) and associated species on the Yukon-Kuskokwim Delta, Alaska. PhD dissertation, University of Michigan, Ann Arbor, MI.

Mickelson, P. G. 1975. *Breeding Biology of Cackling Geese and Associated Species on the Yukon-Kuskokwim Delta, Alaska*. Wildlife Monographs 45. Bethesda, MD: The Wildlife Society.

Mlodinow, S. G., P. Springer, B. Deuel, L. S. Semo, T. Leukering, T. D. Schonewald, W. Tweit, and J. S. Barry. 2008. Distribution and abundance of cackling goose subspecies. *North American Birds* 62: 344–360.

Nelson, U. C., and U. A. Hansen. 1959. The cackling goose—its migration and management. *North American Wildlife Conference Transactions* 24: 174–187.

Petersen, M. R. 1990. Nest-site selection by emperor geese and cackling Canada geese. *Wilson Bulletin* 102: 413–426.

Pierson, B. J., J. M. Pearce, S. L. Talbot, G. F. Shields, and K. T. Scribner. 2000. Molecular genetic analysis of Aleutian Canada geese from Buldir and the Semidi Islands, Alaska. *Condor* 102: 172–180.

Raveling, D. G. 1979. The annual energy cycle in the cackling Canada goose. Pp. 81–93 in: R. L. Jarvis and J. C. Bartonek, eds. *Management and Biology of Pacific Flyway Geese*. Corvallis, OR: OSU Book Stores.

Raveling, D. G. 1984. Geese and hunters of Alaska's Yukon Delta: Management problems and political dilemmas. *Transactions of the North American Wildlife and Natural Resources Conference* 49: 555–575.

Raveling, D. G., J. D. Nichols, J. E. Hines, D. S. Zezulak, J. G. Silveira, J. C. Johnson, T. W. Aldrich, and J. A. Weldon. 1992. Survival of cackling Canada geese, 1982–1988. *Journal of Wildlife Management* 56: 63–73.

Reed, A., P. Dupuis, K. Fisher, and J. Moser. 1980. *An Aerial Survey of Breeding Geese and Other Wildlife in Foxe Basin and Northern Baffin Island, Northwest Territories, July 1979*. Progress Notes 114. Ottawa, ON: Canadian Wildlife Service.

Sedinger, J. S., and D. G. Raveling. 1984. Dietary selectivity in relation to availability and quality of food for goslings of cackling geese. *Auk* 101: 295–306.

Sedinger, J. S., and D. G. Raveling. 1988. Foraging behavior of cackling Canada goose goslings: Implications for the roles of food availability and processing rate. *Oecologia* 75: 119–124.

Sedinger, J. S., and D. G. Raveling. 1990. Parental behavior of cackling Canada geese during brood rearing: Division of labor within pairs. *Condor* 92: 174–181.

Sedinger, J. S., and K. S. Bollinger. 1987. Autumn staging of cackling Canada geese on the Alaska Peninsula. *Wildfowl* 38: 13–18.

Springer, P. F., and R. W. Lowe. 1998. Population, distribution, and ecology of migrating and wintering Aleutian Canada geese. Pp. 425–434 in: D. H. Rusch, et al., eds. *Biology and Management of Canada Geese: Proceedings of the 1998 International Canada Goose Symposium, Milwaukee, Wisconsin*. Madison, WI: Dembar Educational Research Services.

Woolington, D. W., P. F. Springer, and D. R. Yparraguirre. 1979. Migration and wintering distribution of Aleutian Canada geese. Pp. 299–309 in: R. L. Jarvis and J. C. Bartonek, eds. *Management and Biology of Pacific Flyway Geese*. Corvallis, OR: OSU Book Stores.

Barnacle Goose

Black, J. M., and M. Owen. 1988. Variation in pair bond and agonistic behaviors in barnacle geese on the wintering grounds. Pp. 39–57 in: M. W. Weller, ed. *Waterfowl in Winter*. Minneapolis: University of Minnesota Press.

Black, J. M., S. Choudhury, and M. Owen. 1996. Do barnacle geese benefit from long-term monogamy? Pp. 91–117 in: J. M. Black, ed. *Partnerships in Birds: The Study of Monogamy*. Oxford, UK: Oxford University Press.

Madsen, J., and C. E. Mortensen. 1987. Habitat exploitation and interspecific competition in moulting geese in Greenland. *Ibis* 129: 25–44.

Palmer, R. S. 1976. Barnacle goose. Pp. 235–240 in: R. S. Palmer, ed. *Handbook of North American Birds*. Vol. 2: Waterfowl, Part 1. New Haven, CT: Yale University Press.

Ogilvie, M. A., and H. Boyd. 1975. Greenland barnacle geese in the British Isles. *Wildfowl* 26: 139–147.

Owen, M. 1977. *Wildfowl of Europe*. London: Macmillan.

Owen, M. 1980. *Wild Geese of the World: Their Life History and Ecology*. London: B. T. Batsford.

Owen, M., and M. A. Ogilvie. 1979. Wing molt and weights of barnacle geese in Spitsbergen. *Condor* 81: 42–52.

Owen, M., and M. Norderhaug. 1977. Barnacle geese breeding in Svalberg, 1948–1976. *Ornis Scandinavica* 8: 161–174.

Owen, M., and R. H. Kerbes, 1971. On the autumn food of barnacle geese at Caerlaverock National Reserve. *Wildfowl* 22: 114–229.

Reeber, S. 2016. *Waterfowl of North America, Europe and Asia*. Princeton, NJ: Princeton University Press. 656 pp. (Barnacle goose, pp. 259–262)

Salomonsen, F. 1950. *The Birds of Greenland*. Part 1. Copenhagen: Ejnar Munksgaard.

Steen, J. B., and G. W. Gabrielsen. 1986. Thermogenesis in newly hatched eider (*Somateria mollissima*) and long-tailed duck (*Clangula hyemalis*) ducklings and barnacle goose (*Branta leucopsis*) goslings. *Polar Research* 4: 181–186.

Brant

Ankney, C. D. 1984. Nutrient reserve dynamics of breeding and molting brant. *Auk* 101: 361–370.

Anthony, R. M, P. L. Flint, and J. S. Sedinger. 1991. Arctic fox removal improves nest success of black brant. *Wildlife Society Bulletin* 19: 176–184.

Anthony, R. M., W. H. Anderson, J. S. Sedinger, and L. L. McDonald. 1995. Estimating populations of nesting brant using aerial videography. *Wildlife Society Bulletin* 23: 80–87.

Barry, T. W. 1956. Observations on a nesting colony of American brant. *Auk* 73: 193–202.

Barry, T. W. 1962. Effect of late seasons on Atlantic brant reproduction. *Journal of Wildlife Management* 26: 19–26.

Barry, T. W. 1967. Geese of the Anderson River Delta, Northwest Territories, Canada. PhD dissertation, University of Alberta, Edmonton, AB.

Barry, T. W. 1974. Brant, Ross's goose, and emperor goose. Pp. 145–154 in: J. P. Linduska, ed. *Waterfowl Tomorrow*. Washington, DC: US Department of the Interior.

Blurton, P. J. K. 1960. Brent goose population studies, 1958–1959. *Wildfowl Trust Annual Report* 11: 94–98.

Boyd, H., and L. S. Maltby. 1979. The brant of the western Queen Elizabeth Islands, N.W.T. Pp. 5–21 in: R. L. Jarvis and J. C. Bartonek, eds. *Management and Biology of Pacific Flyway Geese*. Corvallis, OR: OSU Book Stores.

Boyd, H., and L. S. Maltby. 1980. Weights and primary growth of brent geese *Branta bernicla* molting in the Queen Elizabeth Islands, N.W.T., Canada, 1973–1975. *Ornis Scandinavica* 11: 135–141.

Boyd, H., L. S. Maltby, and A. Reed. 1988. *Differences in the Plumage Patterns of Brant Breeding in High Arctic Canada*. Progress Notes 174. Ottawa, ON: Canadian Wildlife Service.

Cottam, C., J. J. Lynch, and A. L. Nelson. 1944. Food habits and management of American sea brant. *Journal of Wildlife Management* 8: 36–56.

"C. S. R." 1977. Brent goose. Pp. 436–444 in: S. Cramp and E. L. Simmons, eds. *Handbook of the Birds of Europe, the Middle East, and North Africa: The Birds of the Western Palearctic, Vol. 1. Ostrich to Ducks*. Oxford, UK: Oxford University Press.

Dau, C. P., and D. H. Ward. 1997. *Pacific Brant Wintering at the Izembek National Wildlife Refuge: Population Size, Age Composition, and Breeding Location*. Anchorage, AK: US Fish and Wildlife Service.

Delacour, J., and J. T. Zimmer. 1952. The identity of *Anser nigricans* Lawrence 1846. *Auk* 69: 82–84.

Denny, M. J., H. P. Clausen, S. M. Percival, G. Q. A. Anderson, K. Koffijberg, and J. A. Robinson. 2004. Light-bellied brent goose *Branta bernicla hrota* (East Atlantic Population) in Svalbard, Greenland, Franz Josef Land, Norway, Denmark, the Netherlands and Britain 1960/61–2000/01. *Waterbird Review Series*. The Wildfowl and Wetlands Trust/Joint Nature Conservation Committee, Slimbridge, Gloucester, England, UK.

Derksen, D. V., and D. H. Ward. 1993. *Life History and Habitat Needs of the Black Brant*. Leaflet 13.1.15. Washington, DC: US Fish and Wildlife Service.

Derksen, D. V., K. S. Bollinger, D. H. Ward, J. S. Sedinger, and Y. Miyabayashi. 1996. Black brant from Alaska staging and wintering in Japan. *Condor* 98: 653–657.

Derksen, D. V., W. D. Eldridge, and M. W. Weller. 1982. Habitat ecology of Pacific black brant and other geese moulting near Teshekpuk Lake, Alaska. *Wildfowl* 33: 39–57.

Eichholz, M. W., and J. S. Sedinger. 1998. Factors affecting duration of incubation in black brant. *Condor* 100: 164–168.

Eichholz, M. W., and J. S. Sedinger. 1999. Regulation of incubation behavior in black brant. *Canadian Journal of Zoology* 77: 249–257.

Einarsen, A. S. 1965. *Black Brant: Sea Goose of the Pacific Coast*. Seattle: University of Washington Press.

Flint, P. L., and J. S. Sedinger. 1992. Reproductive implications of egg-size variation in the black brant. *Auk* 109: 896–903.

Handley, C. O., Jr. 1950. The brant of Prince Patrick Island, Northwest Territories. *Wilson Bulletin* 62: 128–132.

Harris, S. W., and P. E. K. Shepherd. 1965. Age determination and notes on the breeding age of black brant. *Journal of Wildlife Management* 29: 643–645.

Hines, J. E., and R. W. Brook. 2008. Changes in annual survival estimates for black brant from the western Canadian arctic, 1962–2001. *Waterbirds* 31: 220–230.

Johnson, S. R. 1993. An important early-autumn staging area for Pacific Flyway brant (*Branta bernicla*): Kasegaluk Lagoon, Chukchi Sea, Alaska. *Journal of Field Ornithology* 64: 539–548.

Jones, R. D., Jr., and D. M. Jones. 1966. The process of family disintegration in black brant. *Wildfowl Trust Annual Report* 17: 75–78.

Kirby, R. E., M. J. Conroy, T. W. Barry, and R. H. Kerbes. 1986. Survival estimates for North American Atlantic brant, 1956–75. *Journal of Wildlife Management* 50: 29–32.

Kramer, G. W., L. R. Rauen, and S. W. Harris. 1979. Populations, hunting mortality, and habitat use of black brant at San Quintín Bay, Baja California, México. Pp. 242–254 in: R. L. Jarvis and J. L. Bartonek, eds. *Management and Biology of Pacific Flyway Geese.* Corvallis, OR: OSU Book Stores.

Ladin, Z. S., P. M. Castelli, S. R. McWilliams, and C. K. Williams. 2011. Time-energy budgets and food use of Atlantic brant across their wintering range. *Journal of Wildlife Management* 75: 273–282.

Lee, D. E., J. M. Black, J. E. Moore, and J. S. Sedinger. 2007. Age-specific stopover ecology of black brant at Humboldt Bay, California. *Wilson Journal of Ornithology* 119: 9–22.

Leopold, A. S., and R. H. Smith. 1953. Numbers and winter distribution of Pacific black brant in North America. *California Fish and Game* 39: 95–101.

Lewis, H. F. 1937. Migration of the American brant. *Auk* 54: 73–95.

Lewis, T. L., P. L. Flint, D. V. Dersken, and J. A. Schmutz. 2011. Fine scale movements and habitat use of black brant during the flightless wing molt in arctic Alaska. *Waterbirds* 34: 177–185.

Lewis, T. L., P. L. Flint, J. A. Schmutz, and D. V. Dersken. 2010. Temporal and spatial shifts in habitat use by black brant immediately following flightless molt. *Wilson Journal of Ornithology* 122: 484–493.

Lindberg, M. S., and J. S. Sedinger. 1997. Ecological consequences of nest site fidelity in black brant. *Condor* 99: 25–38.

Lindberg, M. S., and J. S. Sedinger. 1998. Ecological significance of brood-site fidelity in black brant: spatial, annual, and age-related variation. *Auk* 115: 436–446.

Lindberg, M. S., D. H. Ward, T. L. Tibbitts, and J. Roser. 2007. Winter movement dynamics of black brant. *Journal of Wildlife Management* 71: 534–540.

Lindberg, M. S., J. S. Sedinger, D. V. Derksen, and R. F. Rockwell. 1998. Natal and breeding philopatry in a black brant, *Branta bernicla nigricans*, metapopulation. *Ecology* 79: 1893–1904.

Lindberg, M. S., J. S. Sedinger, and P. L. Flint. 1997. Effects of spring environment on nesting phenology and clutch size of black brant. *Condor* 99: 381–388.

Madsen, J., T. Bregnballe, and F. Mehlum. 1989. Study of the breeding ecology and behaviour of the Svalbard population of light-bellied brent goose *Branta bernicla hrota. Polar Research* 7: 1–21.

Maltby-Prevett, L. S., H. Boyd, and J. D. Heyland. 1975. Observations in Iceland and northwest Europe of brant from the Queen Elizabeth Islands, N.W.T., Canada. *Bird-Banding* 46: 155–161.

Moffitt, J., and C. Cottam. 1941. *Eelgrass Depletion on the Pacific Coast and Its Effect upon Black Brant.* Wildlife Leaflet 204. Washington, DC: US Fish and Wildlife Service.

Moore, J. E., and J. M. Black. 2006a. Historical changes in black brant *Branta bernicla nigricans* use on Humboldt Bay, California. *Wildlife Biology* 12: 151–162.

Morehouse, K. A. 1974. Development, energetics, and nutrition of captive Pacific brant (*Branta bernicla orientalis*). PhD dissertation, University of Alaska Fairbanks.

O'Briain, M., and B. Healy. 1991. Winter distribution of light-bellied brent geese *Branta bernicla hrota* in Ireland. *Ardea* 79: 317–326.

O'Briain, M., A. Reed, and S. D. MacDonald. 1998. Breeding, moulting, and site fidelity of brant (*Branta bernicla*) on Bathurst and Seymour Islands in the Canadian high Arctic. *Arctic* 51: 350–360.

Purcell, J., and A. Brodin. 2007. Factors influencing route choice by avian migrants: a dynamic programming model of Pacific brant migration. *Journal of Theoretical Biology* 249: 804–816.

Raveling, D. G. 1989. Nest-predation rates in relation to colony size of black brant. *Journal of Wildlife Management* 53: 87–90.

Reed, A. 1993. Duration of family bonds of brent geese *Branta bernicla* on the Pacific coast of North America. *Wildfowl* 44: 33–38.

Reed, A. 1997. *Migration Patterns and Philopatry of the Black Brant (*Branta bernicla nigricans*) in the Strait of Georgia, British Columbia.* Technical Report Series 294. Nepean, ON: Canadian Wildlife Service, Pacific and Yukon Region.

Reed, A., E. G. Cooch, F. Cooke, and R. I. Goudie. 1998. Migration patterns of black brant in Boundary Bay, British Columbia. *Journal of Wildlife Management* 62: 1522–1532.

Reed, A., M. A. Davison, and D. K. Kraege. 1989. Segregation of brent geese *Branta bernicla* wintering and staging in Puget Sound and the Strait of Georgia. *Wildfowl* 40: 22–31.

Reed, A., R. Stehn, and D. H. Ward. 1989. Autumn use of Izembek Lagoon, Alaska, by brant from different breeding areas. *Journal of Wildlife Management* 53: 720–725.

Round, P. 1982. Inland feeding by brent geese *Branta bernicla* in Sussex, England. *Biological Conservation* 23: 15–32.

Sedinger, J. S. 1990. Effects of visiting black brant nests on egg and nest survival. *Journal of Wildlife Management* 54: 437–443.

Sedinger, J. S., and C. A. Nicolai. 2011. Recent trends in first-year survival for black brant breeding in southwestern Alaska. *Condor* 113: 511–517.

Sedinger, J. S., and P. L. Flint. 1991. Growth rate is negatively correlated with hatch date in black brant. *Ecology* 72: 496–502.

Sedinger, J. S., C. A. Nicolai, C. J. Lensink, C. Wentworth, and B. Conant. 2007. Black brant harvests, density dependence, and survival: a record of population dynamics. *Journal of Wildlife Management* 71: 496–506.

Sedinger, J. S., C. J. Lensink, D. H. Ward, R. M. Anthony, M. L. Wege, and G. V. Byrd. 1993. Status and recent dynamics of the black brant *Branta bernicla* breeding population. *Wildfowl* 44: 49–59.

Sedinger, J. S., M. P. Herzog, and D. H. Ward. 2004. Early environment and recruitment of black brant (*Branta bernicla nigricans*) into the breeding population. *Auk* 121: 68–73.

Sedinger, J. S., M. S. Lindberg, and N. D. Chelgren. 2001. Age-specific breeding probability in black brant: effects of population density. *Journal of Animal Ecology* 70: 798–807.

Sedinger, J. S., M. S. Lindberg, B. T. Person, M. W. Eichholz, M. P. Herzog, and P. L. Flint. 1998. Density-dependent effects on growth, body size, and clutch size in black brant. *Auk* 115: 613–620.

Sedinger, J. S., M. S. Lindberg, E. A. Rexstad, N. D. Chelgren, and D. H. Ward. 1997. Testing for handling bias in survival estimation for black brant. *Journal of Wildlife Management* 61: 782–791.

Sedinger, J. S., M. W. Eichholz, and P. L. Flint. 1995a. Variation in brood behavior of black brant. *Condor* 97: 107–115.

Sedinger, J. S., N. D. Chelgren, M. S. Lindberg, T. Obritschkewitsch, M. T. Kirk, P. Martin, B. A. Anderson, and D. H. Ward. 2002. Life-history implications of large-scale spatial variation in adult survival of black brant (*Branta bernicla nigricans*). *Auk* 119:510–515.

Sedinger, J. S., P. L. Flint, and M. S. Lindberg. 1995. Environmental influence on life-history traits: growth, survival, and fecundity in black brant (*Branta bernicla*). *Ecology* 76: 2404–2414.

Shields, G. F. 1990. Analysis of mitochondrial DNA of Pacific black brant (*Branta bernicla nigricans*). *Auk* 107: 620–623.

Smith, R. H., and C. H. Jensen. 1970. Black brant on the mainland coast of Mexico. *Transactions of the North American Transactions of the North American Wildlife Conference* 35: 227–241.

Spencer, D. L., U. C. Nelson, and W. A. Elkins. 1951. America's greatest goose-brant nesting area. *Transactions of the North American Wildlife Conference* 16: 290–295.

Taylor, E. J. 1993. Molt and bioenergetics of Pacific black brant (*Branta bernicla nigricans*) on the arctic coastal plain, Alaska. PhD dissertation, Texas A&M University, College Station, TX.

Taylor, E. J. 1995. Molt of black brant (*Branta bernicla nigricans*) on the arctic coastal plain, Alaska. *Auk* 112: 904–919.

Vangilder, L. D., L. M. Smith, and R. K. Lawrence. 1986. Nutrient reserves of premigratory brant during spring. *Auk* 103: 237–241.

Vermeer, K., M. Bentley, K. H. Morgan, and G. E. J. Smith. 1997. Association of feeding flocks of brant and sea ducks with herring spawn at Skidegate Inlet. Pp. 102–107 in: K. Vermeer and K. H. Morgan, eds. *The Ecology, Status, and Conservation of Marine and Shoreline Birds of the Queen Charlotte Islands*. Occasional Paper 93. Ottawa, ON: Canadian Wildlife Service.

Ward, D. H., and P. L. Flint. 1995. Effects of harness-attached transmitters on premigration and reproduction of brant. *Journal of Wildlife Management* 59: 39–46.

Ward, D. H., A. Reed, J. S. Sedinger, J. M. Black, D. V. Derksen, and P. M. Castelli. 2005. North American brant: Effects of changes in habitat and climate on population dynamics. *Global Change Biology* 11: 869–880.

Ward, D. H., C. P. Dau, T. L. Tibbitts, J. S. Sedinger, B. A. Anderson, and J. E. Hines. 2009. Change in abundance of Pacific brant wintering in Alaska: evidence of a climate warming effect? *Arctic* 62: 301–311.

Ward, D. H., D. V. Derksen, S. P. Kharitonov, M. Stishov, and V. V. Baranyuk. 1993. Status of Pacific black brant *Branta bernicla nigricans* on Wrangel Island, Russian Federation. *Wildfowl* 44: 39–48.

Ward, D. H., E. A. Rexstad, J. S. Sedinger, M. S. Lindberg, and N. K. Dawe. 1997. Seasonal and annual survival of adult Pacific brant. *Journal of Wildlife Management* 61: 773–781.

Ward, D. H., J. A. Schmutz, J. S. Sedinger, K. S. Bollinger, P. D. Martin, and B. A. Anderson. 2004. Temporal and geographic variation in survival of juvenile black brant. *Condor* 106: 263–274.

Ward, D. H., R. A. Stehn, and D. V. Derksen. 1994. Response of staging brant to disturbance at the Izembek Lagoon, Alaska. *Wildlife Society Bulletin* 22: 220–228.

Welsh, D. 1988. The relationship of nesting density to behavior and reproductive success of black brant. MS thesis, University of Idaho, Moscow, ID.

Welsh, D., and J. S. Sedinger. 1990. Extra-pair copulations in black brant. *Condor* 92: 242–244.

Wilson, U. W., and J. B. Atkinson. 1995. Black brant winter and spring-staging use at two Washington coastal areas in relation to eelgrass abundance. *Condor* 97: 91–98.

Emperor geese, adults in flight

Lightning Source UK Ltd.
Milton Keynes UK
UKOW05f1123210118
316506UK00001B/3/P